AF368950

Sensors for Vital Signs Monitoring

Sensors for Vital Signs Monitoring

Editors

Jong-Ryul Yang
Eugin Hyun
Sun Kwon Kim

MDPI • Basel • Beijing • Wuhan • Barcelona • Belgrade • Manchester • Tokyo • Cluj • Tianjin

Editors
Jong-Ryul Yang
Electronic Engineering
Yeungnam University
Gyeongsan
Korea, South

Eugin Hyun
Automotive Technology
Daegu Gyeongbuk Institute of
Science and Technology
Daegu
Korea, South

Sun Kwon Kim
Electro-medical Research Center
Korea Electrotechnology
Research Institute
Ansan
Korea, South

Editorial Office
MDPI
St. Alban-Anlage 66
4052 Basel, Switzerland

This is a reprint of articles from the Special Issue published online in the open access journal *Sensors* (ISSN 1424-8220) (available at: www.mdpi.com/journal/sensors/special_issues/Sensors_Vital_Signs_Monitoring).

For citation purposes, cite each article independently as indicated on the article page online and as indicated below:

LastName, A.A.; LastName, B.B.; LastName, C.C. Article Title. *Journal Name* **Year**, *Volume Number*, Page Range.

ISBN 978-3-0365-1766-7 (Hbk)
ISBN 978-3-0365-1765-0 (PDF)

Contents

About the Editors

Jong-Ryul Yang

Prof. Dr. Jong-Ryul Yang was born in Seoul, South Korea, in 1980. He achieved B.S. degrees in Electrical Engineering and Material Science from Ajou University, Suwon, South Korea, in 2003, and a Ph.D. degree in Electrical Engineering from the Korea Advanced Institute of Science and Technology, Daejeon, South Korea, in 2009. From 2009 to 2011, he was a Senior Engineer with the System LSI Division, Samsung Electronics, Yongin, South Korea. From 2011 to 2016, he was a Senior Researcher with the Advanced Medical Device Research Division, KERI, Ansan, South Korea. Since 2016, he has been an Associate Professor with the Department of Electronic Engineering, Yeungnam University, Gyeongsan, South Korea. He is the author of more than 80 journal and conference papers. He holds patents for more than 30 inventions. His research interests include analog/RF/millimeter-wave/terahertz circuits and systems, particularly sub-terahertz imaging systems and miniaturized radar sensors.

Eugin Hyun

Dr. Eugin Hyun was born in Daegu, South Korea, in 1974. He achieved a B.S. degree in Electrical and Electronic Engineering from Yeungnam University, Korea, in 1998, and achieved M.S. and Ph.D. degrees in Electronic Engineering from Yeungnam University, Korea, in 2011 and 2005, respectively. He joined the DGIST (Daegu Gyeongbuk Institute of Science and Technology, Daegu, Korea, in 2005 as a Principal Researcher. From 2007 to 2013, he was also involved in the Department of Electronic Engineering, the Undergraduate Colleges, Yeungnam University, Korea, as an adjunct professor. Additionally, he joined POSTECH (Pohang University of Science and Technology) as a guest researcher in 2018. His primary research interests are radar architecture design, radar signal processing, radar machine learning, and human detection.

Sun Kwon Kim

Dr. Sun Kwon Kim received a Ph.D. degree in Medical Image Processing from the Department of Biomedical Engineering at Seoul National University, Seoul, Korea, in 2011. He was a Research Staff Member at the Samsung Advanced Institute of Technology, Samsung Electronics, Seoul, Korea. Since 2015, he has been a Senior Engineer at the Korea Electrotechnology Research Institute, Ansan, Korea. His main research interest is application development using both software and hardware modifications. He works on the development of AI-portable ophthalmic all-in-one cameras and AI-radar failure correction.

Article

Frontal EEG Changes with the Recovery of Carotid Blood Flow in a Cardiac Arrest Swine Model

Heejin Kim [1], Ki Hong Kim [2], Ki Jeong Hong [2], Yunseo Ku [3,*], Sang Do Shin [2] and Hee Chan Kim [4]

1 Interdisciplinary Program in Bioengineering, Graduate School, Seoul National University, Seoul 03080, Korea; hjkim83@snu.ac.kr
2 Department of Emergency Medicine, Seoul National University Hospital, Seoul 03080, Korea; emphysiciankkh@gmail.com (K.H.K.); emkjhong@gmail.com (K.J.H.); sdshin@snu.ac.kr (S.D.S.)
3 Department of Biomedical Engineering, Chungnam National University College of Medicine, 266, Munwha-ro, Jung-gu, Deajeon 35015, Korea
4 Department of Biomedical Engineering, Seoul National University College of Medicine, Seoul 03080, Korea; hckim@snu.ac.kr
* Correspondence: yunseo.ku@cnu.ac.kr; Tel.: +82-42-280-8613

Received: 17 April 2020; Accepted: 25 May 2020; Published: 28 May 2020

Abstract: Monitoring cerebral circulation during cardiopulmonary resuscitation (CPR) is essential to improve patients' prognosis and quality of life. We assessed the feasibility of non-invasive electroencephalography (EEG) parameters as predictive factors of cerebral resuscitation in a ventricular fibrillation (VF) swine model. After 1 min untreated VF, four cycles of basic life support were performed and the first defibrillation was administered. Sustained return of spontaneous circulation (ROSC) was confirmed if a palpable pulse persisted for 20 min. Otherwise, one cycle of advanced cardiovascular life support (ACLS) and defibrillation were administered immediately. Successfully defibrillated animals were continuously monitored. If sustained ROSC was not achieved, another cycle of ACLS was administered. Non-ROSC was confirmed when sustained ROSC did not occur after 10 ACLS cycles. EEG and hemodynamic parameters were measured during experiments. Data measured for approximately 3 s right before the defibrillation attempts were analyzed to investigate the relationship between the recovery of carotid blood flow (CBF) and non-invasive EEG parameters, including time- and frequency-domain parameters and entropy indices. We found that time-domain magnitude and entropy measures of EEG correlated with the change of CBF. Further studies are warranted to evaluate these EEG parameters as potential markers of cerebral circulation during CPR.

Keywords: cardiopulmonary resuscitation (CPR); electroencephalogram (EEG); hemodynamic data; carotid blood flow (CBF); cerebral circulation

---

1. Introduction

Approximately 395,000 adults experience an out-of-hospital cardiac arrest (OHCA) annually in the US, and their overall survival rate is only 6–11% [1–3]. To prevent death or irreversible damage to vital organs, such as the brain, high-quality cardiopulmonary resuscitation (CPR) is necessary [4,5]. Multiple physiologic measurements have been suggested as indicators of the effectiveness of CPR. End-tidal carbon dioxide (ETCO2) is a widely-used indicator for the pulmonary circulation, and the ETCO2-directed feedback methods are reported to improve the likelihood of return of spontaneous circulation (ROSC) [6].

Recently, achieving good neurological recovery has been regarded as one of the major goals of CPR, because it can influence survivors' quality of life and their socioeconomic burden [7,8]. However,

ETCO2 mainly reflects the systemic circulation, and thus is not adequate to monitor the cerebral circulation or physiological responses of the brain during CPR. Carotid blood flow, the blood supply to the brain, can reflect the cerebral circulation directly. However, its measurement requires an ultrasonic volume flow meter, as well as a skilled operator.

Non-invasive electroencephalography (EEG) can be an alternative to overcome these drawbacks. Portable and low-cost EEG headsets and sensors are currently available out-of-hospital [9]. EEG activity during CPR is reported to be sensitive to cerebral circulation [10,11]. Once cerebral oxygenation decreases due to cardiac arrest (CA), the EEG activity gradually enters the isoelectric state [12–14]. However, the EEG activity can return to the pre-arrest state when ROSC was achieved [15,16]. Effective CPR maintains a certain degree of cerebral electrical activity, changing the EEG activity from isoelectric status to large-amplitude and low-frequency status with bispectral index score (BIS) above 40 [17]. As an important tool for determining the prognosis of ischemic episodes of CA patients, EEG signal is routinely monitored for post-resuscitation treatment [18]. The application of EEG monitoring has expanded to the CPR situation, and distinctive EEG patterns are suggested as possible markers for the quality of cerebral resuscitation and oxygen delivery [19]. To date, however, the direct relationship between the carotid blood flow (CBF) recovery and the EEG during CPR has been rarely discussed.

In this study, we focused on the investigation of the relationship between the recovery of carotid blood flow and non-invasive EEG parameters, including time- and frequency-domain parameters, and entropy indices between defibrillation attempts. We applied a single-channel EEG measurement device that was developed in our laboratory and designed a ventricular fibrillation (VF) swine model with simultaneous measurements of EEG and hemodynamic data, including CBF. We hypothesized that CBF recovery may improve cerebral electrical activity, which can result in EEG changes, even during short intervals between defibrillation attempts.

2. Materials and Methods

2.1. Ethical Statement

The animal test protocol was approved by the Institutional Animal Care and Use Committee of Seoul National University Hospital (IACUC Number: 17-0106). All animal care abided by the Laboratory Animal Act of the Korean ministry of Food and Drug Safety.

2.2. Study Design and Setting

An animal experiment was designed based on a VF swine model. The LUCAS machine (LUCAS2 Chest Compression System, Jolife AB, Sweden) was exploited for mechanical chest compressions (Figure 1a). The machine compressed the chest at a rate of 100 compressions/min with a depth of 5 cm. To prevent displacement of the piston, animals were fixed on the table and the back plate was positioned underneath the animal as a support for the machine. The exact location of the heart was identified by ultrasonic imaging, and then the piston was placed on the site. One emergency medical technician held the machine to prevent the displacement of the piston during CPR.

The assumed scenario of this study was a witnessed OHCA. The duration of untreated VF was determined by considering CA-CALL time (the time of cardiac arrest to call) for bystanders to recognize CA and call emergency services [20]. In addition, it was estimated that four consecutive basic life support (BLS) cycles were performed by the bystanders, with the help of an emergency center, before the dispatched emergency medical team (EMT) arrived at the site. The EMT performed the first defibrillation shock and checked the electrocardiogram (ECG). If the ECG rhythm was shockable, a biphasic defibrillation shock of 200 J was applied by the EMT to restart the heart. Monitoring was initiated when a palpable pulse with organized QRS complexes and systolic blood pressure over 60 mmHg appeared [21]. Sustained ROSC was confirmed if spontaneous circulation continued for 20 min [19]. Once a palpable pulse did not appear after the defibrillation, or VF occurred again during the monitoring session, one cycle of advanced cardiovascular life support (ACLS) was performed by

EMT immediately. If the ECG rhythm was still shockable, the defibrillation shock was applied. In case of pulseless electrical activity or asystole, however, the defibrillation shock was omitted, and the next cycle of ACLS was initiated immediately. If a palpable pulse appeared after the defibrillation, then monitoring for 20 min was initiated. Non-ROSC was confirmed if sustained ROSC was not achieved after 10 cycles of ACLS. During ACLS sessions, epinephrine of 1 mg was injected once every 3 min [22]. After the monitoring sessions or all 10 ACLS sessions, the animals were administered euthanasia, with an injection of potassium chloride. Simultaneous EEG and hemodynamic data were collected during the experiments. The entire test scenario, with a timeline, is described in Figure 2.

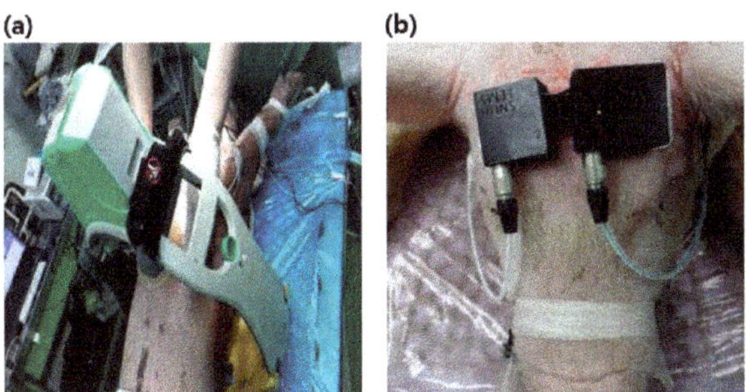

Figure 1. Experimental setup: (**a**) LUCAS2 chest compression system installed on the chest of animals; (**b**) A single-channel electroencephalography (EEG) device mounted on the forehead.

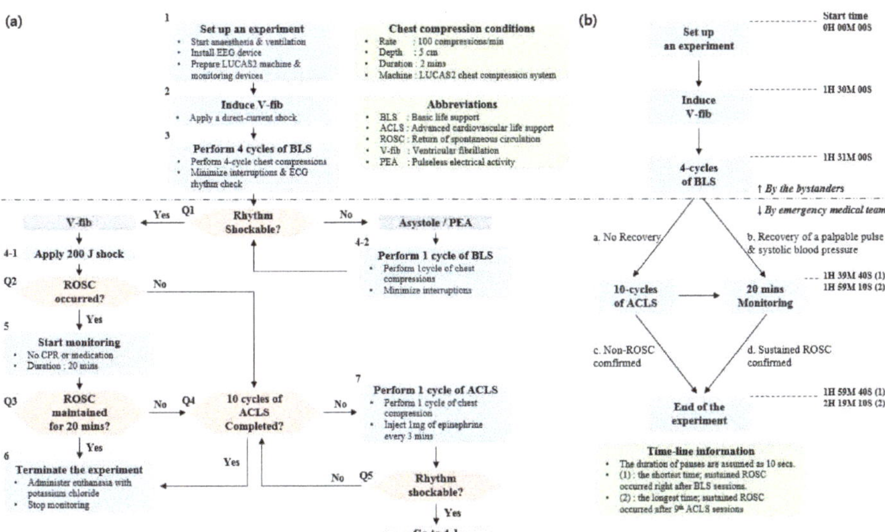

Figure 2. The entire test scenario: (**a**) flow chart from surgical procedure to basic life support (BLS), advanced cardiovascular life support (ACLS), and termination; (**b**) brief timeline of the test protocol.

2.3. Experimental Animals and Housing

Eight domestic cross-bred pigs, approximately 3 months of age (45.6 ± 2.4 kg), were studied. The animals were maintained in an accredited Association for Assessment and Accreditation of Laboratory Animal Care (AAALAC) International (#001169) facility, in accordance with the Guide for the Care and Use of Laboratory Animals [23]. They were judged healthy and fasted overnight.

2.4. Surgical Preparation and Hemodynamic Measurements

The animals were initially sedated with intramuscular injections of 5 mg/kg of tiletamine hydrochloride and zolazepam hydrochloride (Zoletil, Virbac, France) and 2 mg/kg of Xylazine (Rompun, Bayer, Korea), followed by inhaled isoflurane at a dose of 1–1.5%. Endotracheal intubation was performed on the sedated animals, and a capnography (Capstar-100, CWE Inc., Ardmore, PA, USA) was installed. Mechanical ventilation was initiated. To continue the anesthesia, a tidal volume of 12 mL/kg, respiratory rate of 10 breaths/min, partial pressure of arterial carbon dioxide at approximately 40 mmHg, and partial pressure of arterial oxygen over 80 mmHg were maintained.

An implantable perivascular probe (MA2PSB, Transonic Systems, Ithaca, NY, USA) combined with a perivascular flowmeter (T420, Transonic Systems, Ithaca, NY, USA) was placed on the internal carotid artery to measure the CBF. A pressure catheter (Mikro-tip pressure catheter, Millar, Houston, TX, USA) was inserted into the left femoral artery and placed in the descending thoracic aorta to measure the arterial blood pressure. Another Mikro-tip pressure catheter was inserted into the right atrium to measure the right atrial pressure. The ECG and saturation of percutaneous oxygen were also measured. All signals except EEG were gathered and saved in a data acquisition system (PowerLab 16/35, ADInstruments, Dunedin, New Zealand) simultaneously.

A pace-making wire was inserted into the right ventricle through the central vein catheter. Isoflurane was stopped before inducing VF to recover EEG signal. EEG started to recover, and appeared similar to the recording obtained before the injections. Then, a direct-current shock was applied to induce VF. Mechanical ventilation was halted, and the animals were left without assistance for 1 min. Thereafter, CPR and defibrillation attempts were executed, and manual ventilation using a resuscitator bag (Ambu Resuscitators, Ambu A/S, Ballerup, Denmark) was initiated to provide positive pressure ventilation to the animal at a rate of once per 10 compressions.

2.5. EEG Measurement

A portable single-channel digital electroencephalograph and disposable surface electrodes (MT100, Kendall Healthcare, Toronto, Ontario, Canada) were attached to measure the scalp EEG under the referential montage. Reference and ground electrodes were attached on either side of the mastoid. Active electrodes connected to the device were placed on the forehead (Figure 1b). The raw EEG signal was bandpass filtered with a frequency range of 0.5–47 Hz and amplified with a gain of 12,000 v/v. The amplified signal with a low noise level under ±3 µVp-p was digitized and transmitted to the laptop via Bluetooth communication at a rate of 250 Hz. The data acquisition software in the laptop receives and saves the EEG data.

2.6. Data Processing

All data were processed using MATLAB (MATLAB R2017b, Mathworks, Natick, MA, USA). The EEG and hemodynamic data were synchronized. Approximately 3-s-long pauses right before the defibrillation shocks were selected for analysis. The selected EEG was segmented into three 2-s-long sub-epochs with 1.5-s overlaps to reduce variation; 0–2 s, 0.5–2.5 s, and 1–3 s period. The representative EEG parameters were obtained from the average of three sub-epochs. Segmenting the EEG and obtaining parameters is similar to the signal processing technique for the BIS monitor [24]. Time and frequency domain parameters and entropy indices were obtained in this manner. All parameters considered are listed in Table 1.

Table 1. EEG parameters considered in this study.

EEG Parameters	Definition	Domain
Magnitude	Maximal Amplitude during the Epoch (unit: µV)	Time
SynchFastSlow	$\log(B_{0.5-47\,\text{Hz}}/B_{40-47\,\text{Hz}})$	Frequency
BetaR	$\log(P_{30-47\,\text{Hz}}/P_{11-20\,\text{Hz}})$	Frequency
DeltaR	$\log(P_{8-20\,\text{Hz}}/P_{1-4\,\text{Hz}})$	Frequency
AlphaPR	$P_{8-13\,\text{Hz}}/P_{0.5-47\,\text{Hz}}$	Frequency
BetaPR	$P_{13-30\,\text{Hz}}/P_{0.5-47\,\text{Hz}}$	Frequency
DeltaPR	$P_{0.5-4\,\text{Hz}}/P_{0.5-47\,\text{Hz}}$	Frequency
ThetaPR	$P_{4-8\,\text{Hz}}/P_{0.5-47\,\text{Hz}}$	Frequency
BG_Alpha+	$P_{8-47\,\text{Hz}}/P_{0.5-47\,\text{Hz}}$	Frequency
Log energy entropy	$\sum_{i=1}^{n} \log(p(x_i))^2$	Entropy
Rényi entropy	$\frac{1}{1-\alpha} \log(\sum_{i=1}^{n} p(x_i)^{\alpha})$, $(\alpha \geq 0, \neq 1)$	Entropy

Abbreviation: $P_{a-b\,\text{Hz}}$, the sum of spectral power from a–b Hz; $B_{a-b\,\text{Hz}}$, the sum of bispectral activity from a–b Hz; $p(x_i)$, probability distribution function of signal x_i; α of Rényi entropy was 0.5.

2.7. Data Analysis

First, CBF recovery during CPR were analyzed to investigate their relationship with resuscitation rates. The recovery rate was defined as a relative scale of each hemodynamic parameter with respect to the baseline value in the pre-VF state. Second, the EEG waveforms were scrutinized according to the test scenario. EEG activity was evaluated, along with the recovery of CBF.

Pearson correlation coefficients between each EEG parameter and the recovery rates of CBF for all experiments were obtained to inspect whether EEG parameters show similar changes with the CBF. In addition, the recovery rates of CBF were categorized into four quartile groups: group 1 (<25%); group 2 (25–50%); group 3 (50–75%), and group 4 (>75%). Averages of each EEG parameter among groups were evaluated through one-way analysis of variance (ANOVA). Significance was considered at a level of $p < 0.05$. Receiver operating characteristic (ROC) curve analysis was also performed to measure the optimal cut-off values of EEG parameters, to discriminate between the higher and the lower group of the CBF recovery based on the median value, which was approximately 30%. These tests were performed with SPSS (SPSS Statistics 23, IBM SPSS Statistics, New York, NY, USA).

3. Results

3.1. Results of CPR Process

All eight experiments were performed successfully. Once VF was triggered, mean arterial pressure (MAP) decreased to almost 0 mmHg, while approximately 20% of baseline MAP remained as residual pressure in the vessel. CBF dropped rapidly to almost 0% of baseline values during untreated VF. When BLS sessions began, hemodynamic parameters started to recover. Recovery rates of hemodynamic parameters over the BLS and ACLS sessions are presented specifically in Supporting File S1.

Sustained ROSC was achieved in five animals. Among them, one animal was defibrillated after the last BLS session. Another four animals were defibrillated during the course of the ACLS sessions. No animals experienced VF again during the monitoring sessions. Three animals were not resuscitated until the tenth ACLS session was completed. BLS cycles were performed a total of 32 times, and ACLS cycles were performed 48 times, and data after those sessions were included for analysis.

3.2. EEG Changes with the Recovery of CBF

The EEG waveforms between an ROSC (Test 6) and a non-ROSC (Test 5) case were compared (Figure 3). Before VF, the amplitude of EEG with irregular morphology exceeding ±20 µV was observed.

Since cerebral oxygenation decreased due to VF, the amplitude started to decrease in 10–15 s and almost entered the isoelectric state (±5 μV) at the end of untreated VF.

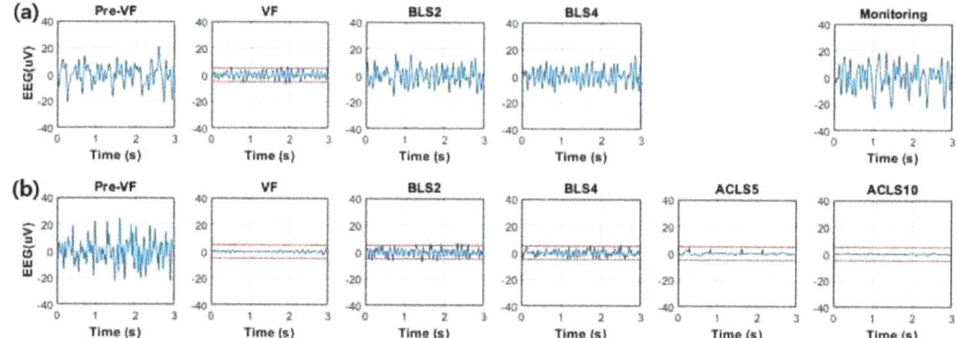

Figure 3. Comparison of EEG over time between return of spontaneous circulation (ROSC) and non-ROSC cases: (**a**) EEG waveforms obtained from animals with successful defibrillation after fourth BLS session and sustained ROSC confirmed after follow-up monitoring for 20 min (Test 6); (**b**) EEG waveforms obtained from animals in which ROSC was not achieved until the end of experiment (Test 5). Dashed lines denote the level of ±5 μV, the limits of the isoelectric state.

The recovery of the EEG was different, depending on the recovery of the CBF. In Test 6, which showed a better recovery, the recovery rate reached almost 40% during the last two BLS sessions. Concurrently, an increased background activity with higher amplitude and increased higher frequency components was observed. EEG activity during the monitoring session appeared similar to the baseline values during the pre-VF period. This means that the cerebral circulation was restored successfully, whereas the recovery rates in Test 5 exceeded 30% during the first BLS session but decreased consistently during the rest of the CPR sessions. The EEG decreased in amplitude and entered the suppression status and increased lower frequency components during the second BLS session. The cerebral resuscitation was poor, with the low CBF recovery rates of below 10%. Nearly flat patterns resulting from electrocerebral inactivity appeared, and EEG did not recover until the end of the ACLS sessions.

Table 2 shows the Pearson correlation coefficients between EEG parameters and the recovery rates of CBF. Among them, time-domain magnitude and two entropy indices, log energy entropy [25] and Rényi entropy [26], showed a correlation coefficient of approximately 0.78. Figure 4 demonstrates the scatter plots for these three parameters.

Table 2. Pearson correlation coefficients between EEG parameters and the recovery rates of CBF.

EEG Parameters	Correlation Coefficient	*p*-Value
Magnitude	0.778	<0.001
SynchFastSlow	0.210	0.228
BetaR	−0.329	0.016
DeltaR	0.196	0.032
AlphaPR	0.189	0.048
BetaPR	0.323	0.001
DeltaPR	0.032	0.797
ThetaPR	−0.354	0.004
BG_Alpha+	0.262	0.006
Log energy entropy	0.781	<0.001
Rényi entropy	0.784	<0.001

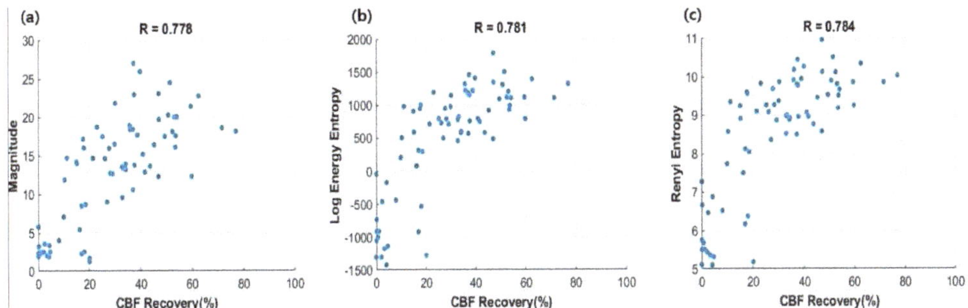

Figure 4. Scatter plots between EEG parameters and the recovery of CBF. Correlation coefficients were denoted above the plots: (**a**) magnitude; (**b**) log energy entropy; (**c**) Rényi entropy.

3.3. Changes in EEG Parameters Depending on Four CBF Groups

Figure 5 illustrates the results of one-way ANOVA tests for three parameters. For magnitude, the lowest quartile (group 1) showed significant differences to other groups, with $p < 0.05$. However, significant difference was not confirmed among the other three groups. Similar patterns were observed in following two entropy indices. Table 3 demonstrates the results of the post hoc test based on the Dunnett T3 method.

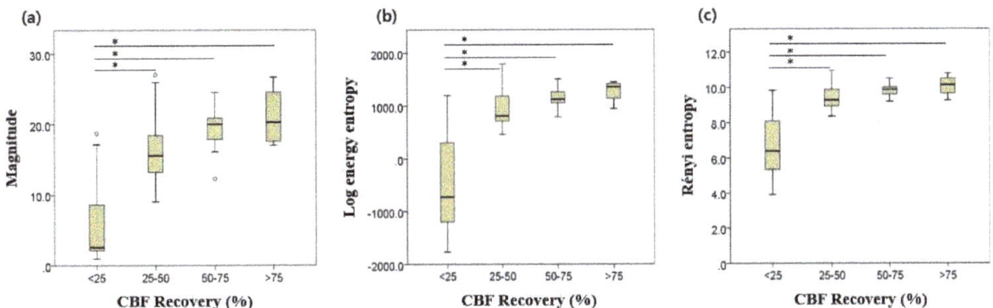

Figure 5. Results of one-way ANOVA: (**a**) magnitude; (**b**) log energy entropy; (**c**) Rényi entropy. Asterisk (*) denotes statistical significance at the $p < 0.001$ level. Error bars indicate the upper and lower extreme values of the data.

Table 3. Results of multiple comparisons between groups in three EEG parameters.

Group I/ Group II		Magnitude	Log Energy Entropy	Rényi Entropy
		Mean Difference/Standard Deviation (*p*-value)	Mean Difference/Standard Deviation (*p*-value)	Mean Difference/ Standard Deviation (*p*-value)
1	2	−10.39/1.24 (<0.001)	−1375.15/164.65 (<0.001)	−2.69/0.319 (<0.001)
1	3	−13.34/1.37 (<0.001)	−1590.42/164.87 (<0.001)	−3.13/0.321 (<0.001)
1	4	−15.15/2.39 (0.012)	−1720.80/190.02 (<0.001)	−3.38/0.434 (<0.001)
2	3	−2.95/1.30 (0.169)	−215.27/87.24 (0.108)	−0.442/0.171 (0.084)
2	4	−4.75/2.35 (0.395)	−345.65/128.60 (0.180)	−0.695/0.338 (0.384)
3	4	−1.80/2.41 (0.958)	−130.39/128.87 (0.871)	−0.253/0.340 (0.957)

Differences were obtained by Group I minus Group II.

3.4. EEG Parameters Depending on the Different CBF Recovery Groups

Figure 6 illustrates the ROC curves for the three EEG parameters. All possible cut-off values are plotted with a combination of true positive rate (sensitivity) and false positive rate (1 − specificity). The optimal cut-off points are also denoted. Table 4 presents the results of ROC curve analysis including area under the curve (AUC), true positive rate, false positive rate, and cut-off values. The AUC values of all three parameters were over 0.88.

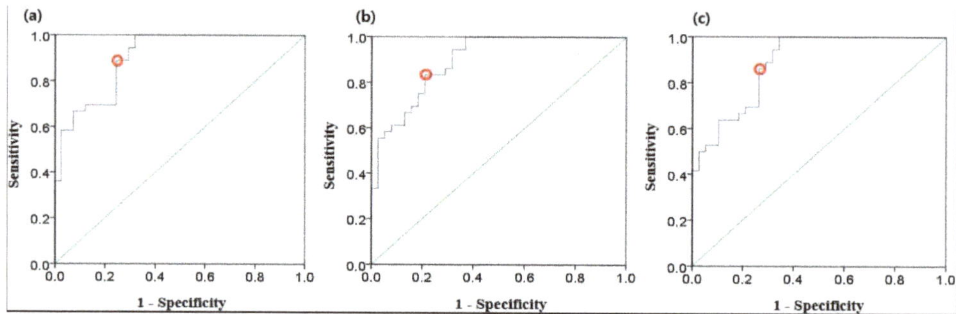

Figure 6. Receiver operating characteristic (ROC) curves (blue) for three EEG parameters: (**a**) magnitude; (**b**) log energy entropy; (**c**) Rényi entropy. Red dots indicate the optimal cut-off points, and the diagonal lines (green) indicate random chance.

Table 4. Results of the ROC curve analysis for EEG parameters.

EEG Parameter	AUC	Standard Error	True Positive Rate (Sensitivity)	False Positive Rate (1-Specificity)	Cut-off Value
Magnitude	0.904	0.033	0.889	0.244	12.802
Log energy entropy	0.896	0.035	0.833	0.211	739.543
Rényi entropy	0.885	0.037	0.861	0.263	8.919

Abbreviation AUC: Area under the curve.

4. Discussion

This study investigated the relationship between the EEG and CBF, to evaluate the feasibility of non-invasive EEG parameters as potential predictors of the recovery of CBF in the CA swine model. The current CPR protocol consists of an ECG rhythm check, chest compression (CC), defibrillation, and medication [22], while CBF or EEG measurement and analysis have rarely been performed during CPR. Monitoring cerebral circulation could provide beneficial information to improve patients' prognosis and quality of life [7,8]. EEG was considered as one of the possible markers because it could reflect the level of cerebral circulation [27]. Post-resuscitation care could be seriously disrupted with a sparsity of EEG activity [28]. If the EEG could reflect the CBF and be measurable in the OHCA setting, CPR with a feedback of non-invasive EEG parameters could guide EMTs to achieve a higher CBF recovery, for example, by guiding leg elevation or the Trendelenburg position [29], which is expected to improve brain perfusion and neurologic outcomes of CA patients after CPR. It is noteworthy that the present study used only single-channel EEG signals from forehead sites where the installation of EEG sensors is convenient.

Several studies have attempted to apply the BIS monitor during CPR. However, unwanted artefacts due to CCs contaminated the original EEG, and generated unreliable outputs [30,31]. The BIS monitor is not adequate to use for the short intervals between CCs, because it is based on the moving-average function over 60 s [24]. Prolonged no- or low-flow periods can deteriorate brain function of CA patients [19,32]. Thus, this study focused on data measured during short pauses between the defibrillation attempts. We observed that the EEG background activity increased and became more irregular with the CBF recovery. The frequency distribution of EEG was also affected. As the CBF recovered, the higher frequency components including alpha (8–13 Hz) and beta (13–30 Hz) increased, whereas the lower frequency components including delta (<4 Hz) and theta (4–8 Hz) decreased. These changes affected the functional dynamics associated with varying amplitudes and multi-frequency responses, including the level of complexity and the amount of energy, diversity, and randomness [33], which could be indicated by the increase of log energy entropy and Rényi entropy, as shown in Figure 4. Entropy parameters have been applied to EEG signals, especially in anesthesia or epileptic seizure studies [34–36]. A previous study analyzing epileptic EEG signals reported that log energy entropy of the modulated EEG signals obtained from the epileptogenic area had relatively lower values [35]. Consistently, another study showed that the complexity derived by Rényi entropy was also higher in healthy signals [36]. These parameters might also have a potential to identify sufficient cerebral circulation for satisfying the metabolic requirements of brain cells of CA patients [27].

This study has several limitations. First, the experimental model was finalized assuming a witnessed OHCA. EEG parameters, such as log energy entropy and Rényi entropy, might be able to reflect the cerebral resuscitation only with very short no- or low-flow duration (<1 min). The association between the CBF and EEG recovery is probably less pronounced with a longer untreated VF. Further research should be performed to validate this method with a longer VF period for at least 5 min. Second, this study was performed only with the limited sample size of eight animals. Feature analysis with larger datasets should be performed to confirm our findings. To generalize our findings to real OHCA patients, moreover, future clinical studies should be guaranteed with the experimental setup optimized for human anatomy.

5. Conclusions

We measured a single-channel EEG non-invasively during CPR and evaluated the relationship between EEG parameters and the CBF recovery. Our findings indicated that time-domain magnitude and entropy indices of EEG, even during the brief pause in CPR, may correlate with the level of cerebral circulation. Further studies are warranted to evaluate these parameters as potential markers of cerebral resuscitation.

Supplementary Materials: The following are available online at http://www.mdpi.com/1424-8220/20/11/3052/s1, Figure S1. Experimental setup; Figure S2. The entire test scenario; Figure S3. Comparison of EEG over time between ROSC and non-ROSC cases; Figure S4. Scatter plots between EEG parameters and the recovery of CBF. Correlation coefficients were denoted above the plots; Figure S5. Results of one-way ANOVA; Figure S6. ROC curves (blue) for three EEG parameters; Table S1. EEG parameters considered in this study; Table S2. Pearson correlation coefficients between EEG parameters and the recovery rates of CBF; Table S3. Results of multiple comparisons between groups in three EEG parameters; Table S4. Results of the ROC curve analysis for EEG parameters; Supporting File S1. Hemodynamic changes of individual animals throughout the experiments.

Author Contributions: H.K. conceived of the presented idea, analyzed the data and wrote the manuscript. Y.K. supervised the findings of this work and contributed to the final version of the manuscript. K.H.K. and K.J.H. carried out the animal experiments. S.D.S. and H.C.K. contributed to the design and implementation of the research. All authors discussed the results and approved the final manuscript.

Funding: This study was supported by the Encouragement Program for the Industries of Economics Cooperation Region, funded by South Korea's ministry of Trade, Industry and Energy (Grant Number: R0004491), and by the research fund of Chungnam National University.

Conflicts of Interest: The authors declare no conflict of interest.

References

1. Benjamin, E.J.; Blaha, M.J.; Chiuve, S.E.; Cushman, M.; Das, S.R.; Deo, R.; Floyd, J.; Fornage, M.; Gillespie, C.; Isasi, C. Heart disease and stroke statistics-2017 update: A report from the American Heart Association. *Circulation* **2017**, *135*, e146–e603. [CrossRef] [PubMed]
2. Chan, P.S.; McNally, B.; Tang, F.; Kellermann, A. Recent trends in survival from out-of-hospital cardiac arrest in the United States. *Circulation* **2014**, *130*, 1876–1882. [CrossRef] [PubMed]

3. Daya, M.R.; Schmicker, R.H.; Zive, D.M.; Rea, T.D.; Nichol, G.; Buick, J.E.; Brooks, S.; Christenson, J.; MacPhee, R.; Craig, A. Out-of-hospital cardiac arrest survival improving over time: Results from the Resuscitation Outcomes Consortium (ROC). *Resuscitation* **2015**, *91*, 108–115. [CrossRef] [PubMed]

4. Cummins, R.O.; Ornato, J.P.; Thies, W.H.; Pepe, P.E. Improving survival from sudden cardiac arrest: The "chain of survival" concept. A statement for health professionals from the Advanced Cardiac Life Support Subcommittee and the Emergency Cardiac Care Committee, American Heart Association. *Circulation* **1991**, *83*, 1832–1847. [CrossRef] [PubMed]

5. Iwami, T.; Nichol, G.; Hiraide, A.; Hayashi, Y.; Nishiuchi, T.; Kajino, K.; Morita, H.; Yukioka, H.; Ikeuchi, H.; Sugimoto, H.; et al. Continuous Improvements in "Chain of Survival" Increased Survival After Out-of-Hospital Cardiac Arrests. *Circulation* **2009**, *119*, 728–734. [CrossRef]

6. Hartmann, S.M.; Farris, R.W.; Di Gennaro, J.L.; Roberts, J.S. Systematic review and meta-analysis of end-tidal carbon dioxide values associated with return of spontaneous circulation during cardiopulmonary resuscitation. *J. Intensive Care Med.* **2015**, *30*, 426–435. [CrossRef]

7. Moulaert, V.R.; Verbunt, J.A.; van Heugten, C.M.; Wade, D.T. Cognitive impairments in survivors of out-of-hospital cardiac arrest: A systematic review. *Resuscitation* **2009**, *80*, 297–305. [CrossRef]

8. Torgersen, J.; Strand, K.; Bjelland, T.; Klepstad, P.; Kvåle, R.; Søreide, E.; Wentzel-Larsen, T.; Flaatten, H. Cognitive dysfunction and health-related quality of life after a cardiac arrest and therapeutic hypothermia. *Acta Anaesthesiol. Scand.* **2010**, *54*, 721–728. [CrossRef]

9. Berka, C.; Levendowski, D.J.; Cvetinovic, M.M.; Petrovic, M.M.; Davis, G.; Lumicao, M.N.; Zivkovic, V.T.; Popovic, M.V.; Olmstead, R. Real-Time Analysis of EEG Indexes of Alertness, Cognition, and Memory Acquired With a Wireless EEG Headset. *Int. J. Hum.–Comput. Interact.* **2004**, *17*, 151–170. [CrossRef]

10. Moss, J.; Rockoff, M. EEG monitoring during cardiac arrest and resuscitation. *JAMA* **1980**, *244*, 2750–2751. [CrossRef]

11. England, M.R. The changes in bispectral index during a hypovolemic cardiac arrest. *Anesthesiology* **1999**, *91*, 1947. [CrossRef] [PubMed]

12. Aminoff, M.J.; Scheinman, M.M.; Griffin, J.C.; Herre, J.M. Electrocerebral accompaniments of syncope associated with malignant ventricular arrhythmias. *Ann. Intern. Med.* **1988**, *108*, 791–796. [CrossRef]

13. De Vries, J.W.; Bakker, P.F.; Visser, G.H.; Diephuis, J.C.; van Huffelen, A.C. Changes in cerebral oxygen uptake and cerebral electrical activity during defibrillation threshold testing. *Anesth. Analg.* **1998**, *87*, 16–20. [PubMed]

14. Pana, R.; Hornby, L.; Shemie, S.; Dhanani, S.; Teitelbaum, J. Time to loss of brain function and activity during circulatory arrest. *J. Crit. Care* **2016**, *34*, 77–83. [CrossRef] [PubMed]

15. Losasso, T.J.; Muzzi, D.A.; Meyer, F.B.; Sharbrough, F.W. Electroencephalographic monitoring of cerebral function during asystole and successful cardiopulmonary resuscitation. *Anesth. Analg.* **1992**, *75*, 1021–1024. [CrossRef] [PubMed]

16. Schäuble, B.; Klass, D.W.; Westmoreland, B.F. Resuscitation artifacts during electroencephalography. *Am. J. Electroneurodiagn. Technol.* **2002**, *42*, 16–21. [CrossRef]

17. Azim, N.; Wang, C. The use of bispectral index during a cardiopulmonary arrest: A potential predictor of cerebral perfusion. *Anaesthesia* **2004**, *59*, 610–612. [CrossRef]

18. Feng, G.; Jiang, G.; Li, Z.; Wang, X. Prognostic value of electroencephalography (EEG) for brain injury after cardiopulmonary resuscitation. *Neurol. Sci.* **2016**, *37*, 843–849. [CrossRef]

19. Reagan, E.M.; Nguyen, R.T.; Ravishankar, S.T.; Chabra, V.; Fuentes, B.; Spiegel, R.; Parnia, S. Monitoring the Relationship Between Changes in Cerebral Oxygenation and Electroencephalography Patterns During Cardiopulmonary Resuscitation: A Feasibility Study. *Crit. Care Med.* **2018**, *46*, 757–763. [CrossRef]

20. Mullie, A.; Lewi, P.; Van Hoeyweghen, R.; Group, C.R.S. Pre-CPR conditions and the final outcome of CPR. *Resuscitation* **1989**, *17*, S11–S21. [CrossRef]

21. Wei, L.; Yu, T.; Gao, P.; Quan, W.; Li, Y. Detection of Spontaneous Pulse Using Acceleration Signals Acquired From CPR Feedback Sensor in Porcine Model of Cardiac Arrest. *Circulation* **2016**, *134* (Suppl. 1), A12827. [CrossRef] [PubMed]

22. Peberdy, M.A.; Callaway, C.W.; Neumar, R.W.; Geocadin, R.G.; Zimmerman, J.L.; Donnino, M.; Gabrielli, A.; Silvers, S.M.; Zaritsky, A.L.; Merchant, R. Part 9: Post–cardiac arrest care: 2010 American Heart Association guidelines for cardiopulmonary resuscitation and emergency cardiovascular care. *Circulation* **2010**, *122* (Suppl. 3), S768. [CrossRef] [PubMed]

23. Council, N.R. *Guide for the Care and Use of Laboratory Animals*; National Academies Press: Washington, DC, USA, 2010.

24. Rampil, I.J. A primer for EEG signal processing in Anesthesia. *Anesthesiol. Phila. Hagerstown* **1998**, *89*, 980–1002. [CrossRef] [PubMed]

25. Raghu, S.; Sriraam, N.; Kumar, G.P. Classification of epileptic seizures using wavelet packet log energy and norm entropies with recurrent Elman neural network classifier. *Cogn. Neurodyn.* **2017**, *11*, 51–66. [CrossRef]

26. Acharya, U.R.; Fujita, H.; Sudarshan, V.K.; Bhat, S.; Koh, J.E. Application of entropies for automated diagnosis of epilepsy using EEG signals: A review. *Knowl.-Based Syst.* **2015**, *88*, 85–96. [CrossRef]

27. Foreman, B.; Claassen, J. Quantitative EEG for the detection of brain ischemia. *Crit. Care* **2012**, *16*, 216. [CrossRef]

28. Bhalala, U.; Polglase, G.; Dempsey, E. Editorial: Neonatal and Pediatric Cerebro-Cardio-Pulmonary Resuscitation (CCPR): Where Do We Stand and Where Are We Heading? *Front. Pediatr.* **2018**, *6*, 165. [CrossRef]

29. Zadini, F.; Newton, E.; Abdi, A.A.; Lenker, J.; Zadini, G.; Henderson, S.O. Use of the trendelenburg position in the porcine model improves carotid flow during cardiopulmonary resuscitation. *West. J. Emerg. Med.* **2008**, *9*, 206–211.

30. Chollet-Xémard, C.; Combes, X.; Soupizet, F.; Jabre, P.; Penet, C.; Bertrand, C.; Margenet, A.; Marty, J. Bispectral index monitoring is useless during cardiac arrest patients' resuscitation. *Resuscitation* **2009**, *80*, 213–216. [CrossRef]

31. Fatovich, D.M.; Jacobs, I.G.; Celenza, A.; Paech, M.J. An observational study of bispectral index monitoring for out of hospital cardiac arrest. *Resuscitation* **2006**, *69*, 207–212. [CrossRef]

32. Sekhon, M.S.; Ainslie, P.N.; Griesdale, D.E. Clinical pathophysiology of hypoxic ischemic brain injury after cardiac arrest: A "two-hit" model. *Crit. Care* **2017**, *21*, 90. [CrossRef] [PubMed]

33. Rosso, O.A.; Blanco, S.; Yordanova, J.; Kolev, V.; Figliola, A.; Schürmann, M.; Başar, E. Wavelet entropy: A new tool for analysis of short duration brain electrical signals. *J. Neurosci. Methods* **2001**, *105*, 65–75. [CrossRef]

34. Liang, Z.; Wang, Y.; Sun, X.; Li, D.; Voss, L.J.; Sleigh, J.W.; Hagihira, S.; Li, X. EEG entropy measures in anesthesia. *Front. Comput. Neurosci.* **2015**, *9*, 16. [CrossRef] [PubMed]

35. Das, A.B.; Bhuiyan, M.I.H. Discrimination and classification of focal and non-focal EEG signals using entropy-based features in the EMD-DWT domain. *Biomed. Signal Process. Control* **2016**, *29*, 11–21. [CrossRef]

36. Yin, Y.; Sun, K.; He, S. Multiscale permutation Rényi entropy and its application for EEG signals. *PLoS ONE* **2018**, *13*. [CrossRef] [PubMed]

Article

A Patient-Specific 3D+t Coronary Artery Motion Modeling Method Using Hierarchical Deformation with Electrocardiogram

Siyeop Yoon [1,2], **Changhwan Yoon** [3], **Eun Ju Chun** [4] and **Deukhee Lee** [1,2,*]

1   Center for Medical Robotics, Korea Institute of Science and Technology, 5, Hwarang-ro 14-gil, Seongbuk-gu, Seoul 02792, Korea; h14515@kist.re.kr
2   Division of Bio-medical Science and Technology, KIST School, Korea University of Science and Technology, Seoul 02792, Korea
3   Cardiovascular Center, Seoul National University Bundang Hospital, Seongnam 13620, Korea; kunson2@snu.ac.kr
4   Department of Radiology, Seoul National University Bundang Hospital, Seongnam 13620, Korea; humandr@snubh.org
*   Correspondence: dkylee@kist.re.kr; Tel.: +82-10-958-5633

Received: 3 September 2020; Accepted: 30 September 2020; Published: 5 October 2020

Abstract: Cardiovascular-related diseases are one of the leading causes of death worldwide. An understanding of heart movement based on images plays a vital role in assisting postoperative procedures and processes. In particular, if shape information can be provided in real-time using electrocardiogram (ECG) signal information, the corresponding heart movement information can be used for cardiovascular analysis and imaging guides during surgery. In this paper, we propose a 3D+t cardiac coronary artery model which is rendered in real-time, according to the ECG signal, where hierarchical cage-based deformation modeling is used to generate the mesh deformation used during the procedure. We match the blood vessel's lumen obtained from the ECG-gated 3D+t CT angiography taken at multiple cardiac phases, in order to derive the optimal deformation. Splines for 3D deformation control points are used to continuously represent the obtained deformation in the multi-view, according to the ECG signal. To verify the proposed method, we compare the manually segmented lumen and the results of the proposed method for eight patients. The average distance and dice coefficient between the two models were 0.543 mm and 0.735, respectively. The required time for registration of the 3D coronary artery model was 23.53 s/model. The rendering speed to derive the model, after generating the 3D+t model, was faster than 120 FPS.

Keywords: 3D+t modeling; coronary artery; non-rigid registration; cage deformation; 4D CT

1. Introduction

Cardiovascular disease (CVD) is one of the primary causes of death worldwide, with 22.2 million deaths expected by 2030. According to National Health and Nutrition Examination Survey (NHANES) data from 2013 to 2016, the prevalence of CVD was 48.0% in adults over the age of 20. The prevalence of CVD has a positive correlation with an increase with age [1]. The resulting social cost is estimated to have been 351.3 billion dollars in the U.S. alone, from 2014 to 2015. In particular, cardiovascular disease accounted for 14% of total medical spending in U.S., the highest rate among other major diagnostic groups—even higher than cancer. In the global population, the burden of expenditure is even more serious [1].

Providing sufficient information through image analysis acquired in the pre-operative diagnosis stage eliminates unnecessary examination and helps in developing patient-specific treatment plans.

As the heart is a continuously beating organ and there may be unexpected movements in patients (e.g., arrhythmia), if information on this movement can be obtained in advance, coronary artery and heart procedures may be more efficient.

As shown in Figure 1, the ECG signal is a change in potential that is correlated with the movement of the heart muscle. Motion of the heart produces changes in volume and pressure in the cardiac chambers; therefore, ECG provides important information about the movement of the heart. When CT is reconstructed by performing retrospective ECG synchronization, the movements of the heart and coronary arteries (according to the cardiac cycle) can be obtained geometrically, and movement information (e.g., coronary artery distortion), as well as characteristics of the coronary stenosis, can be obtained.

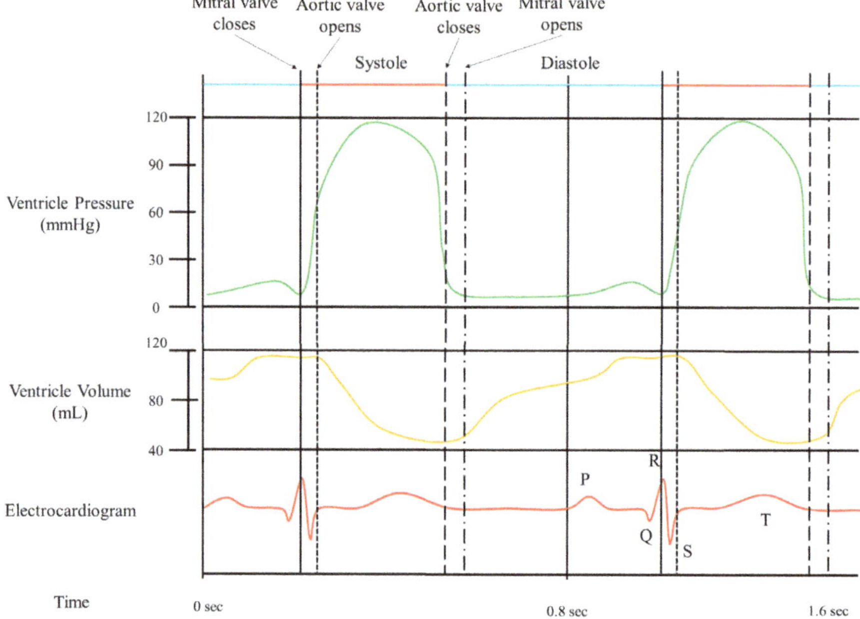

Figure 1. The ECG signal and volume of the ventricle during the different phases of a cardiac cycle.

Patient-specific 4D heart shape information facilitates the following applications: accurate coronary artery structure acquisition [2,3], analysis of 4D blood flow and stenosis [4–6], removal of motion artifacts in the vascular region [7–9], surgical simulation for each patient [10], postoperative evaluation and analysis [5,6], atrial motion analysis [11,12], vascular motion analysis [13,14], and real-time deformation prediction during surgery combined with 2D images [8].

In particular, cardiac CT angiography (CTA) has an isotropic spatial resolution of less than 0.5 mm and, so, can be used to observe the movement of the coronary artery and trabeculae of the ventricle. In addition to grading the degree of calcification of the coronary artery and the total amount of plaque from the CT image, it is also possible to measure the torsion of the coronary artery at a high resolution [15]. This high-resolution spatial information can help the operator perform a procedure appropriate for each patient before and after surgery. However, despite the high spatial resolution of the CTA, its temporal resolution is 50–200 ms, which is lower than that of 4D echocardiography (30–100 Hz) and cardiac MRI (30–50 ms). Therefore, proper shape interpolation for restoring high time resolution information from CTA imagery is essential for co-operation with the other applications while preserving high-precision anatomical details.

One of the essential requirements of cardiac modeling in many of these applications is that the topology of the mesh model constituting the cardiac model must be consistently preserved. In particular, in the case of a 4D heart model, the mesh should not cause new problems (e.g., self-intersection or mesh degeneration), even if the positions of the vertices constituting the shape change over time. To address these requirements, many researchers have employed the template-based registration scheme.

The registration process matches a template model to a target model through geometric shape deformation. The most popular registration algorithm is the iterative closest point (ICP) method, which consists of finding the correspondence between two models and finding the optimal transformation [16]. However, the ICP method is sensitive to the initial position and noise, while the shape registration is limited up to rigid transformation. Non-rigid registration deals with the deformation of the shape in addition to rigid transformations. Non-rigid registration is more challenging, however, as non-rigid transformations not only require more correspondences to be defined, but the solution space is much more extensive [17]. Research into the registration of 3D non-rigid shapes has been actively conducted using the methods of delineating shape deformation and shape correspondence.

To describe the deformation of an object, with respect to its dynamics and material properties, many researchers have assumed shape deformation to be a physical model, such as a linear elastic model [18], non-linear elastic model [19,20], viscous fluid [21], or diffusion model [22,23]. In particular, the Large Deformation Diffeomorphic Metric Mapping (LDDMM) framework provides robust deformation as a massive flow consisting of diffeomorphisms [24]. However, the physical models are computationally expensive and sensitive to mechanical properties. On the other hand, the statistical shape deformation model (SSM) uses a low-dimensional statistical model, in which shape deformation is inferred from a training data set [25,26]. Although SSM reduces the computational cost, the shape of variability is limited by the training data. Therefore, the deformation is hardly representative of the inter-variability of patients, such as the topological discontinuity of coronary arteries. Nora et al. described the motion modeling problem using the coronary arteries attached to the SSM of muscles [27]. Due to the representation of the coronary artery, the deformation poorly described the lumen diameter. Instead of modeling a priori physical and statistical information, there have been attempts to estimate the shape transformations with landmarks and coherent motions. Radial basis function methods express shape deformations as weighted sums of distance function for control point changes [28]. In particular, the thin-plate spline method minimizes the bending energy, which has a closed-form solution [29]. The most popular form of deformation is known as B-spline free-form deformation (FFD) [26,30,31]. Rueckert et al. [32] have proved the conditions for FFD to have diffeomorphic deformation. This method has disadvantages, however: the numerical cost increases with the number of control points and the degree of freedom of shape deformation is fixed. Therefore, the deformation has limited representation capacity.

In addition to deformation modeling, establishing correspondences between shapes is a critical problem in registration. The one-to-one correspondence of the ICP method is sensitive to the initial position and shape loss. To determine many-to-many correspondence points, Chui et al. [33] used the fuzzy correspondence between two shapes. The problem of selecting a robust point matching the correspondence has been interpreted as a combination of a Gaussian mixture model (GMM) and Expectation Maximization [34]. In the GMM model, one point is the centroid of the Gaussian distribution for the points constituting the shape, while the other point is regarded as the data to generate [35]. The variations of GMM have different deformation models, according to the obtained transformation parameters and the regularization term. For example, regularizing the second derivative of the transformation leads to a thin-plate spline transformation, while regularizing according to motion coherence theory leads to a coherent point drift transformation [36,37]. The variations of GMM have been generalized using the generalized Gaussian radial basis function [38].

To represent the local spatial representation, an L_2E estimator has been proposed, which creates a robust sparse–dense correspondence [39,40].

However, the mixture model requires a high computational cost, as it generates a Gaussian distribution for each of the points. Furthermore, each Gaussian distribution shares its standard deviation among the points. The mixture model is thus sensitive to noise and shape loss. Especially for coronary arteries, narrow and tangled structures are very challenging to model in 3D+t. This is because the loss of blood vessel morphology can be observed in different heartbeats of the same patient, through motion artifacts and geometrical deformations.

In this paper, we propose a 3D+t coronary artery model that can be inferred in real-time, according to ECG signals. The overall structure of the proposed method is shown in Figure 2. The proposed patient-specific 3D+t coronary artery motion model is divided into two processing blocks, according to the timing of data processing; (1) a preoperative processing block, and (2) an intraoperative usage block.

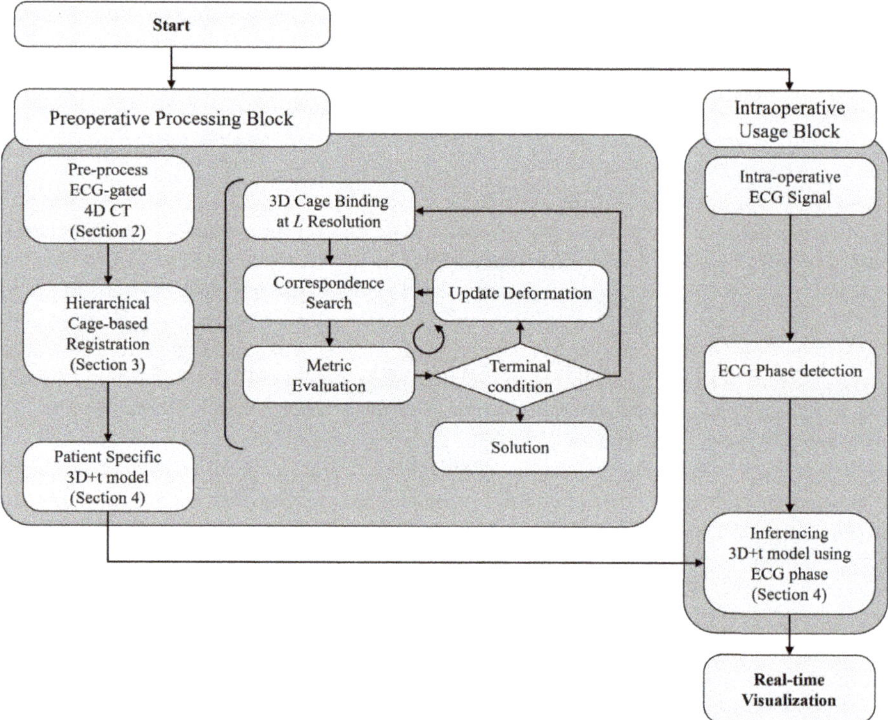

Figure 2. General framework of the proposed method.

At the preprocessing step, we first perform the segmentation of 4D CTA volumes to generate artery models. After we have multiple coronary artery models, hierarchical cage-based registration is performed to construct a patient-specific 3D+t model with hyperelastic regularization. The proposed hierarchical cage deformation model more robustly/accurately registers coronary artery models in different cardiac phases. When updating the control points of the coronary artery model, we gradually increase the degrees of freedom of the deformation model. A modified hyper-elastic regularization term prevents mesh degeneration problems during the control point optimization step. After the optimal cage control point is obtained—which minimizes the shape dissimilarity of the source shape and the target shape—we interpolate the shape control point to build a continuous 3D+t model. The interpolated shape model provides fast shape-inference for intraoperative usage.

In the intraoperative usage step, the ECG phase can be assessed from the patient's real-time signal, which is then correlated to the geometric deformation of cardiac muscles, as shown in Figure 1. Therefore, the 3D shape of the coronary artery is provided by a continuous 3D+t model, according to change of the ECG signal and time.

The contributions of the proposed method consist of the following:

1. A hierarchical deformation method to perform robust shape registration, even with incomplete coronary artery models;
2. Rapid shape interpolation that enables restoring small and complex geometry in a time-varying coronary artery model;
3. The modified hyper-elastic regularization prevents mesh degeneration during shape registration; and
4. Evaluation of the proposed method using retrospective data for eight patients, both qualitatively and quantitatively.

2. Pre-Processing ECG-Gated 4D CT Images

In this study, we reconstructed CTA volumes for eight patients at 0% to 95% intervals (in 5% intervals) between the RR peaks of the heart rate. We took the volumes using a 256-slice multi-detector CT scanner (BRILLIANCE ICT 256 SLICE, Philips Healthcare) at the Cardiovascular Center of Seoul National University Bundang Hospital. This retrospective study was approved by the Institutional Review Board of Seoul National University Bundang Hospital (IRB No. B-2009-637-103).

The cross-sectional size of the image was 512×512 pixels, the average number of slices in the Z-axis direction was 298, and the volume voxel resolution was $0.35 \times 0.35 \times 0.45$ mm. The left ascending and circumflex coronary arteries were segmented using the ITK-Snap software [41]. The segmented arteries were converted to mesh models using Poisson surface reconstruction, where the average number of nodes was 10,638. We selected 75% phase mesh models as templates, as the left ascending and circumflex coronary arteries are most clearly observed at 75% phase [42]. Figure 3 shows the models of the template and other phases.

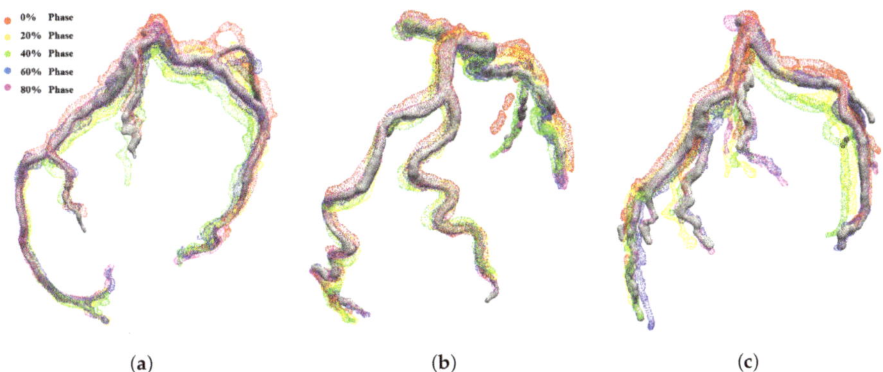

(a) (b) (c)

Figure 3. Left ascending and circumflex coronary arteries of: (a) patient 1; (b) patient 2; and (c) patient 4. The template phase model is a white solid model, while the other phases are colored with respect to their cardiac phases.

3. Hierarchical Cage-Based Shape Registration Method

In this section, we address the non-rigid registration method to find the optimal deformation between coronary artery models. This section is organized as follows: (1) shape representation and registration problems; (2) gradient descent for shape control point optimization; (3) multi-resolution cage deformation representation; and (4) diffeomorphism supported by hyper-elasticity regularization;

3.1. Shape Representation and Registration Problems

This section provides the basic concepts of representation and registration of shapes. Let a shape V be $V = \{v_i | v_i \in \mathbb{R}^3, i = 0, \dots, n-1\}$, which contains n of vertices. If V_s and V_t are source and target shapes, respectively, the registration problem is to find an optimal transformation that minimizes the dissimilarity between the shapes. Here, an arbitrary transformation T maps the source shape V_s to the target shape V_t. Through the optimization process, the optimal transformation parameters x^* minimize disparity measure as follows:

$$x^* = \arg \min_x d(T(x) \circ V_s, V_t). \tag{1}$$

The transformation T is a mapping such that

$$T : V_s \rightarrow \bar{V} = V_s + U(V_s, x), \tag{2}$$

where $\bar{V}$ is the deformed shape, x are the local deformation parameters, and $U(V, x)$ is a vertex-wise mapping.

The shape transformation T may be represented through the modification of a coarse cage mesh that envelops the source shape. Let a region Ω bound the shape V_s in 3D. The sub-region Ω_r is a sub-divisions of Ω, where $\Omega = \bigcup_{\forall r} \Omega_r$ and $\Omega_i \cap \Omega_j = \emptyset, i \neq j$. If we create the $m \times m \times m$ regular lattice grid on the region Ω, the sub-divisions of Ω contain $(m-1)^3$ control vertices and m^3 sub-regions. Hereby, the sub-region Ω_r is defined as an 8-point cuboid. The eight corner points of Ω_r are given as $P_r = \{p_i | p_i \in \mathbb{R}^3, i = 0, \dots, 7\}$. The linear combination of cage control points and their local coordinates represent the vertices of the shape V_s. If the vertex $v \subset \Omega_r$, then the representation of the vertex by the sub-region control point is given as below:

$$v = F(v; P_r) = \sum_{i=0}^{7} \varphi_i(v) p_i, \tag{3}$$

where ϕ_i is a trilinear shape function for assigning local coordinates, such as

$$\begin{aligned}
\varphi_0(v_x, v_y, v_z) &= (1 - v_x)(1 - v_y)(1 - v_z)/8 \\
\varphi_1(v_x, v_y, v_z) &= (1 + v_x)(1 - v_y)(1 - v_z)/8 \\
\varphi_2(v_x, v_y, v_z) &= (1 + v_x)(1 + v_y)(1 - v_z)/8 \\
\varphi_3(v_x, v_y, v_z) &= (1 - v_x)(1 + v_y)(1 - v_z)/8 \\
\varphi_4(v_x, v_y, v_z) &= (1 - v_x)(1 - v_y)(1 + v_z)/8 \\
\varphi_5(v_x, v_y, v_z) &= (1 + v_x)(1 - v_y)(1 + v_z)/8 \\
\varphi_6(v_x, v_y, v_z) &= (1 + v_x)(1 + v_y)(1 + v_z)/8 \\
\varphi_7(v_x, v_y, v_z) &= (1 - v_x)(1 + v_y)(1 + v_z)/8.
\end{aligned} \tag{4}$$

From the previous definition of a cage representation, the motion of vertex v in the direction $\bar{v}$ is given as follows:

$$\begin{aligned}
u(v, P_r) &= \bar{v} - v \\
&= \sum_{i=0}^{7} \varphi_i(v)(p_i + \partial p_i) - \sum_{i=0}^{7} \varphi_i(v) p_i \\
&= \sum_{i=0}^{7} \varphi_i(v) \partial p_i,
\end{aligned}$$

where $u(v, x)$ is the motion of vertex v and ∂p_i is the motion of control point p_i. Therefore, the shape deformation is only dependent on the change of control points, as shown in Figure 4. Therefore, the parameter of the cage representation of the transformation T is given as follows:

$$x = \{\partial p_i | \partial p_i \in \mathbb{R}^3, i = 0, \dots, 7\}. \tag{5}$$

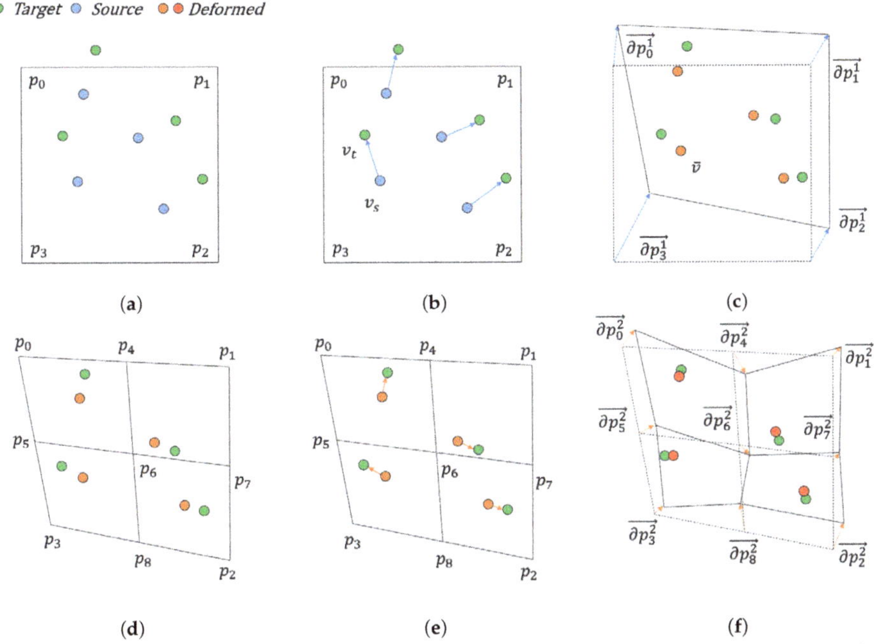

Figure 4. Hierarchical registration with different deformation depths. (**a–c**) Level 1; (**d–f**) Level 2 registration. (**a,d**) show cage partitioning at different levels. (**b,e**) show correspondence searching, while (**c,e**) show the gradient descent-based deformation update.

3.2. Gradient Descent for Shape Control Point Optimization

The optimization of the transformation is defined as the process of minimizing a metric. If we set the disparity measure as the squared Euclidean distance between the correspondence pair, then

$$d(v_s, v_s, P) = \|T(P) \circ v_s - v_t\|^2 \tag{6}$$

$$= \|F(v_s; P) - v_t\|^2 \tag{7}$$

$$= \|\sum_{i=0}^{7} \varphi_i(v_s) p_i - v_t\|^2, \tag{8}$$

where $v_s \in V_s$, $v_t \in V_t$, and v_s, v_t is correspondence pair.

Thus, the gradient can be denoted as the sum of the difference vector of the corresponding pair multiplied by the weight of each control point. Therefore, the partial derivative of the disparity measure for a cage control point $p_i = (p_{ix}, p_{iy}, p_{iz})$ is

$$\frac{\partial}{\partial p_i} d(v_s, v_t, P) = \frac{\partial}{\partial p_i} \|v_s - v_t\|^2 \tag{9}$$

$$= 2\|v_s - v_t\| \cdot \frac{\partial}{\partial p_i} \left\| \sum_{i=0}^{7} \varphi_i(v_s) p_i - v_t \right\| \tag{10}$$

$$= 2\|v_s - v_t\| \cdot \varphi_i(v_s). \tag{11}$$

The Jacobian matrix for the disparity measure of one correspondence pair is

$$\frac{\partial}{\partial P} d(v_s, v_t, P) = \begin{bmatrix} \frac{\partial}{\partial p_0} d(v_s, v_t, P) \\ \frac{\partial}{\partial p_1} d(v_s, v_t, P) \\ \vdots \\ \frac{\partial}{\partial p_m} d(v_s, v_t, P) \end{bmatrix} = \begin{bmatrix} \partial p_{0x} & \partial p_{0y} & \partial p_{0z} \\ \partial p_{1x} & \partial p_{1y} & \partial p_{1z} \\ & \vdots & \\ \partial p_{mx} & \partial p_{my} & \partial p_{mz} \end{bmatrix}. \tag{12}$$

The update of cage control points uses the distribution of the differences of corresponding pairs, which are the vectors from sources to targets generated inside the sub-regions Ω_r. At this time, the robust correspondence selection potentially supports the update of cage control points. To establish the correspondence pair robustly, we constrain the correspondence searching process using orientation filtering, as follows:

$$\{v_s, v_t\} = \begin{cases} \text{Paired,} & \text{if } \theta(\vec{n}_s, \vec{n}_t) < \theta_{\text{Threshold}} \\ \text{Not paired,} & \text{otherwise} \end{cases}, \tag{13}$$

where $\theta(\cdot, \cdot)$ is angle between the two vectors, and $\vec{n}_s$ and $\vec{n}_t$ are the vertex normals of v_s and v_t, respectively. We set $\theta_{\text{Threshold}} = 30°$. The Jacobian matrix for the sum of the squared Euclidean distance is:

$$\sum_{\{v_s, v_t\} \in \forall I_{V_s}} \frac{\partial}{\partial P} d(v_s, v_t, P) = \begin{bmatrix} \sum \partial p_0 \\ \sum \partial p_1 \\ \vdots \\ \sum \partial p_m \end{bmatrix}, \tag{14}$$

where I_{V_s} is the set of correspondence pairs.

3.3. Multi-Resolution Cage Deformation Representation

In this section, we present a cage deformation method using multi-resolution to represent a gradual deformation. In the registration process, the resolution of the cage determines the degree of freedom of shape deformation. With increasing resolution of the cage, the deformation model can represent a more detailed shape change. However, a dense cage has the disadvantage that it can lose the overall shape. A method for maintaining local shape features through multi-resolution or hierarchical data structures is, thus, used as a complementary method.

We assumed the generation of the cage based on a regular lattice grid. The primitive shape of the cage obtained from the lattice structure is a cube with eight vertices and six quadrilateral faces, where the points inside the cage can be represented as linear combinations of cage control vertices. As shown in Figure 4, the cage can be partitioned into the inner sub-regions, where the control points of this sub-region can be created using the control points of the outer region. We denote the vertex v using cage control points P^n at the deformation depth of n as follows:

$$v = F(v; P^n) = \sum_{i=0}^{7} \varphi_i(v) p_i^n, \tag{15}$$

where p_i^n is ith cage control point at the deformation depth n. If we recursively acquire the sub-region of the region Ω that surrounds the source model, we can denote the higher-level control points using the lower level control points. The generalized formula presents the corner points of the sub-division, which are recursively described in the multi-resolution process below:

$$P^m = F(P^m; P^n) = \sum_{i=0}^{7} \varphi_i(P^m) p_i^n, \tag{16}$$

where $m > n$ and $m, n \in \mathbb{N}$. Thus, if we represent the shape using a chain of cage deformations, the deformed vertex $\tilde{v}$, with respect to the level n deformation, is

$$\tilde{v} = F(v; P^n, \partial P^n) = \sum_{i=0}^{7} \varphi_i(v)(p_i^n + \partial p_i^n). \tag{17}$$

Similarly, the deformed vertex $\tilde{v}$ by level n deformation after level $n - 1$ deformation is

$$\tilde{v} = F(v; P^n, P^{n-1}) \tag{18}$$
$$= \sum_{i=0}^{7} \varphi_i(v)(\sum_{j=0}^{7} \varphi_j(p_i)(p_j^{n-1} + \partial p_j^{n-1}) + \partial p_i^n).$$

To cooperate with the gradient descent, we reformulate $\frac{\partial}{\partial p_i} d(v_s, v_t, P)$ as a multi-resolution process. The partial derivative of the given cost function at the $(n-1)^{\text{th}}$ level is given as follows:

$$\frac{\partial}{\partial p_i^{n-1}} d(v_s, v_t, P) = \frac{\partial}{\partial p_i^{n-1}} \| \sum_{i=0}^{7} \varphi_i(v_s) p_i^n - v_t \|^2$$
$$= \frac{\partial}{\partial p_i^{n-1}} \| \sum_{i=0}^{7} \varphi_i(v_s) \sum_{j=0}^{7} \varphi_j(p_i^n) p_j^{n-1} - v_t \|^2$$
$$= 2\|v_s - v_t\| \cdot \| \sum_{i=0}^{7} \varphi_i(v_s) \varphi_j(p_i^n) \|. \tag{19}$$

Modification of the control point ∂p_i in the multi-resolution cage sub-division is carried out by

$$\partial p_i = \partial p_i^n + \partial p_i^{n-1} + \cdots + \partial p_i^1. \tag{20}$$

3.4. Diffeomorphism Supported by Hyper-Elasticity Regularization

Although hierarchical cage deformation recursively represents shape deformation to avoid local minima, dense cages possibly lead to more cage degeneration. Therefore, an appropriate regularization process is required when applying hierarchical transformations. For plausible deformation, we used hyper-elastic regularization, which prevents unexpected partial deformation. We utilized and modified the study of Burger et al. [20], which can be easily extended to the cage deformation setting. As shown in Figure 5, the 24 sub-regions of the cage were defined using corner points p_i and seven auxiliary points, which are the volume points p_V and face points p_F. Tetrahedral sub-regions are defined by the span of a volume point and corresponding face points. Regularization ensures that the transformation is a diffeomorphism; that is, it is reversible and smooth. Hyper-elastic regularization, as defined by Burger et al. [20], is given by

$$S^{hyper}(x) = \int \alpha_1 \eta_{vol}(x) + \alpha_2 \eta_{sur}(x) + \alpha_3 \eta_{len}(x) d\Omega. \tag{21}$$

where α_i are balancing parameters. The functions η_{vol}, η_{sur}, and η_{len} penalize changes of volume, surface, and length, respectively. Here, we set the balancing parameter as 10.0 for all experiments. Burger et al. [20] utilized the average points to delineate the volume point p_V and six face points p_F. However, if the cage is concave (i.e., due to large deformations), the face and volume points are not maintained inside the cage, as shown in Figure 5. As a result, the functions η_{vol} and η_{sur} may have negative values, which can lead to the failure of gradient descent.

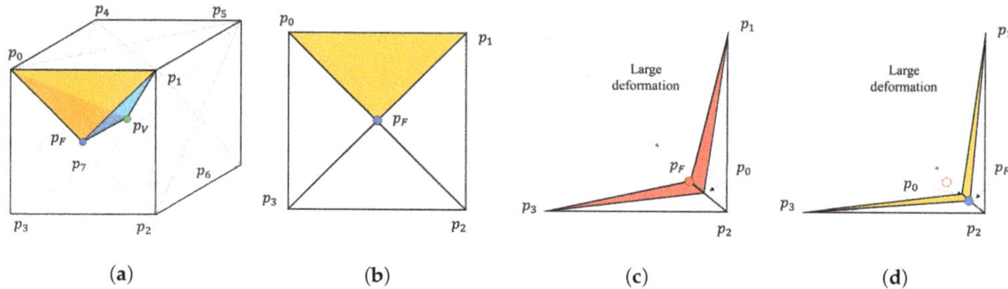

(a)   (b)   (c)   (d)

Figure 5. Cage sub-division with hyper-elastic regularization: (**a**) A tetrahedral sub-division of the 3D cage volume, which is the span of face (blue) and volume (green) points; (**b**) sub-division of cage face; (**c**) the average points (red) located outside of the cage and their negative areas (red triangles); and (**d**) the equal-area points (blue) located inside of cages despite large deformations and their positive areas (yellow triangles).

To achieve robust regularization, we define the face and volume vertices of each cage to have the same sub-area and sub-volume inside of the cage. Assuming that the face point $p_F = (p_{F_x}, p_{F_y}, p_{F_z})$ is located inside the quadrilateral, the position p_F of the points dividing the areas $\triangle p_0 p_1 p_F$, $\triangle p_1 p_2 p_F$, $\triangle p_2 p_3 p_F$, and $\triangle p_3 p_0 p_F$ is defined as follows;

$$\triangle p_i p_{i+1} p_F = (p_{i+1} - p_i) \times (p_F - p_i)/2 = [p_{i+1} - p_i] \times (p_F - p_i)/2, \tag{22}$$

where $i = \{0, 1, 2, 3\}$. The least-squares solution of the above conditions for all triangles is

$$\begin{bmatrix} [p_1 - p_0]_\times \\ [p_2 - p_1]_\times \\ [p_3 - p_2]_\times \\ [p_0 - p_3]_\times \end{bmatrix} \begin{bmatrix} p_{F_x} \\ p_{F_y} \\ p_{F_z} \end{bmatrix} = \begin{bmatrix} [p_1] \times p_0 \\ [p_2] \times p_1 \\ [p_3] \times p_2 \\ [p_1 0] \times p_3 \end{bmatrix}. \tag{23}$$

Similar to the face point, we assume that the volume point is located inside the hexahedron. The volume point $p_V = (p_{V_x}, p_{V_y}, p_{V_z})$ partitions 24 sub-tetrahedra of the cage. The volume of a single tetrahedron is given as

$$V_{p_{i,j} p_{i+1,j} p_{F_j}} = (p_{i+1,j} - p_{i,j}) \times (p_{F_j} - p_{i,j}) \cdot (p_V - p_{i,j})/6, \tag{24}$$

where p_{F_j} is jth face point of the hexahedron and $p_{i,j}$ is the ith corner point of the jth face.

The volume point p_V is obtained by solving the following least-squares problem:

$$\begin{bmatrix} [p_{1,0} - p_{0,0}] \times p_{F_0} - [p_{1,0}] \times p_{0,0} \\ [p_{2,0} - p_{1,0}] \times p_{F_0} - [p_{2,0}] \times p_{1,0} \\ [p_{3,0} - p_{2,0}] \times p_{F_0} - [p_{3,0}] \times p_{2,0} \\ [p_{0,0} - p_{3,0}] \times p_{F_0} - [p_{0,0}] \times p_{3,0} \\ \vdots \\ [p_{0,5} - p_{3,5}] \times p_{F_5} - [p_{0,5}] \times p_{3,5} \end{bmatrix} \begin{bmatrix} p_{V_x} \\ p_{V_y} \\ p_{V_z} \end{bmatrix} = \begin{bmatrix} -[p_{1,0}] \times p_{0,0} \cdot p_{F_0} \\ -[p_{2,0}] \times p_{1,0} \cdot p_{F_0} \\ -[p_{3,0}] \times p_{2,0} \cdot p_{F_0} \\ -[p_{0,0}] \times p_{3,0} \cdot p_{F_0} \\ \vdots \\ -[p_{0,5}] \times p_{3,5} \cdot p_{F_5} \end{bmatrix}. \tag{25}$$

The robust face/volume points improve the numerical stability of the cage deformation. The cost function is a combination of the dissimilarity measurement and regularization functions.

4. Interpolation of Shape Control Points

In this section, we introduce the shape interpolation and restoration for real-time usage of the 3D+t coronary artery model. According to Equation 3, the shape of the coronary artery relies on the locations of the control points. Therefore, as we derive the intermediate positions of control points among cardiac phases, the corresponding shape is restored. To interpolate the positions of control points, we consider a set of control point at the kth phase as a vector P_k, such that $P_k = \{p_0, p_1, \ldots, p_{n-2}, p_{n-1}\}$, where $p_i \in \mathbb{R}^3$ and n is the number of cage control points. From the registration results, we interpolate the given sets of control points using periodic cubic spline interpolation [43], due to the (cyclic) nature of the heart's motion. The number of knots is the same as the number of reconstructions from 4D CTA.

Let a phase-varying vector $S(t) = \{s_0(t), \ldots, s_{n-1}(t)\}$ be the set of interpolated control points, where $s_i(t)$ is the ith control spline for the cardiac phase t. The spline vector $S(t)$ has C^2 continuity with respect to the phase t. The spline function $S(t)$ maps phase t to the set of cage control points, such that $S : \mathbb{R} \to \mathbb{R}^{3 \times n}$. Then, the vertices of shape are restored using the following equation:

$$v(t) = F(v; S(t)) = \sum_{i \in I_V} \varphi_i(v) s_i(t). \tag{26}$$

5. Evaluations and Results

We evaluated the proposed method both qualitatively and quantitatively on data from eight patients. The proposed method was tested on an Intel (R) Xeon (R) W-2133 workstation with CPU@3.60 GHz and 32 GB ram. We partially multi-threaded the computation of cost function measurements using OpenMp [44] and Thread building block [45] during the optimization process. The proposed method and comparison target methods were written in C++.

5.1. Quantitative Evaluations

In the quantitative evaluation of non-rigid registration, we used metrics considering: (1) the closest point-mesh Euclidean distance (ED) from the target model to the matching result and (2) the dice coefficient (DC) obtained from the mesh boolean operation. As we set the number of iterations to 300/maxDepth for each depth, the total number of iterations for different max depths was set to be the same.

5.1.1. Trade-off between Deformation Depth and Computation Time

First, we observed the trade-off between the degrees of freedom of deformation and computation time. As shown in Table 1, we compared the different depths of deformation incrementally, from 1 to 5. As shown in Figure 6, the ED and DC worsened, as the phases were far from the template phase. As the shape of the blood vessel was a thin tube shape, the ED and DC values noticeably deteriorated with slight movement. As the deformation depth increased, the ED values gradually decreased and the DC values increased more prominently; both metrics flattened for the other cardiac phases. The metrics converged after deformation depth 4. The comparison results for the other patients are given in Appendix A.

Table 1. Trade-off between computation time and accuracy.

Method	Cage Resolution	Computation Time (s)	Average Distance (mm)	Dice Coefficient
HierCage	[1, 1, 1]	21.73	0.668 ± 0.255	0.655 ± 0.096
HierCage	[2, 2, 2]	23.05	0.597 ± 0.234	0.696 ± 0.077
HierCage	[3, 3, 3]	22.91	0.566 ± 0.227	0.721 ± 0.068
HierCage	*[4, 4, 4]*	*23.52*	*0.543 ± 0.222*	*0.735 ± 0.064*
HierCage	[5, 5, 5]	33.00	0.534 ± 0.221	0.741 ± 0.064
GRBF_KC	[4, 4, 4]	40.99	0.615 ± 0.218	0.666 ± 0.088
GRBF_L2	[4, 4, 4]	40.92	0.600 ± 0.207	0.671 ± 0.084
TPS_KC	[4, 4, 4]	33.00	0.553 ± 0.191	0.681 ± 0.080
TPS_L2	[4, 4, 4]	32.21	0.530 ± 0.17	0.690 ± 0.075

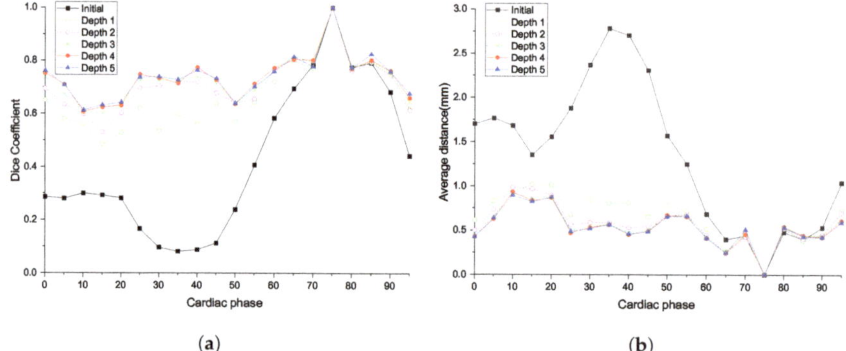

Figure 6. Effect of cage deformation depth for patient 1 in different cardiac phases: (**a**) dice coefficients; and (**b**) average distance from target to deformed model.

5.1.2. Comparison with Other Methods

In the second experiment, the proposed registration method's performance was compared with that of other non-rigid matching algorithms. As the comparison target of non-rigid registration, we selected the variations of GMM methods, which are combinations of a deformation model and a cost function. The deformation models were thin-plate spline (TPS) and generalized radial basis function (GRBF), while the cost functions were kernel correlation (KC) or L_2 distance. The comparison targets used L-BFGS-B as an optimization method.

For comparison, each deformation model had the same number of deformation control points. Considering the convergence of accuracy from the previous analysis, we set the number of grid and control points as $16 \times 16 \times 16$ and 4913, respectively. As shown in Figure 7, the proposed method had higher DCs than the comparison targets at the interval [15%, 35%] of cardiac phases, where the interval had a DC value of less than 0.2 before registration. Although the ED metrics showed a similar trend, compared with other methods, a significant improvement in DC was observed. The comparison results for the other patients are given in Appendix A.

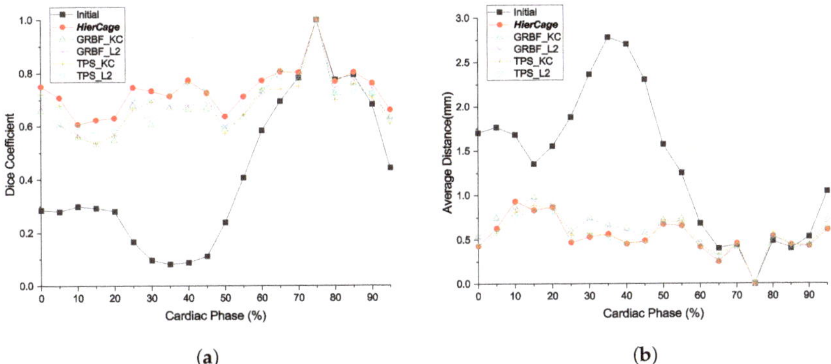

(a)   (b)

Figure 7. Comparison with other algorithms for patient 1 at the different cardiac phases: (**a**) dice coefficients; and (**b**) average distance from target to deformed model.

5.1.3. Interpolation Accuracy

Furthermore, we evaluated the accuracy of the interpolated 3D+t coronary artery models by comparing them with the segmented models. The proposed method created a smooth and non-degenerate 3D model by interpolating the cage control points over sampled cardiac phases, as shown in Figure 8. Our data sets were evenly reconstructed from 4D CT within the R-R peak with 5% sampling interval. Thus, we had 20 keyframes ($V_{0\%}$, $V_{5\%}$, ..., $V_{95\%}$). To evaluate the effect of sampling the cardiac phases, we chose phase sets from the given 20 keyframes as follows: (1) Odd 10: [5, 15, 25, ..., 85, 95]; (2) Even 10: [0, 10, 20, ..., 80, 90]; (3) Odd 5: [5, 25, 45, 65, 85]; and (4) Even 5: [0, 20, 40, 60, 80]. Figure 9 shows the differences among the phase selections. Although the sampled phase became sparse, the DC of the interpolated result showed that the results had a lower bound. The ED still showed a flattened value, when compared to that before registration. The comparison results for the other patients are given in Appendix A.

(a)   (b)   (c)   (d)

Figure 8. Comparison of the registered model and interpolated model for patient 5: (**a**) Template (green) and 40% coronary artery (red); (**b**) registered model (blue); (**c**) interpolated model (yellow); and (**d**) comparison of registered model and interpolated model.

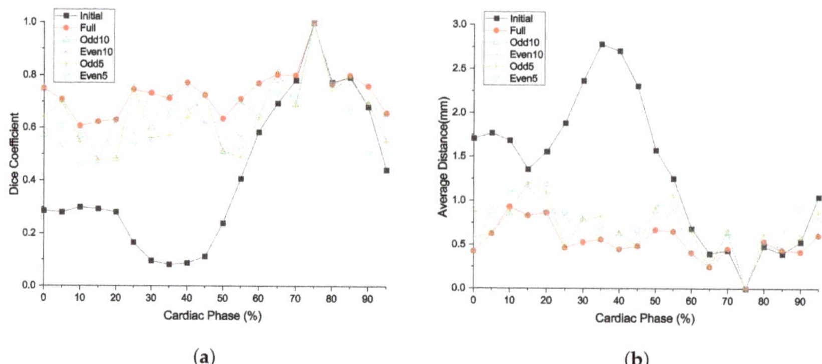

(a)   **(b)**

Figure 9. Comparison of interpolation sampling for patient 1 in the different cardiac phases: (**a**) dice coefficients; and (**b**) average distance from target to deformed model.

5.2. Qualitative Evaluations

In the qualitative evaluation, the role of the visualization effect of hyper-elastic regularization, geometrical comparison with other algorithms, comparison of the matching result and interpolation result model, and the limitations of the interpolation model were investigated.

5.2.1. The Effect of Hyper-Elastic Regularization and Hierarchical Deformation

High-order deformation models often converge to a local minimum, which may look visually implausible. Figure 10a,b show examples of shape shrinkage when the target model contains loss of shape. The hyper-elastic regularization constraints lead to shape preservation, thus providing a plausible result, as shown in Figure 10c,d.

When compared with the other algorithms, Figure 11 shows an example where shape registration is defective at the excessively deformed and twisted parts. As the registration process converged to a local minimum, the deformation model represents the further details of local deformation. On the other hand, hierarchical cage deformation gradually acquired an optimal solution, passing from coarse to dense resolution, to avoid local minima, as shown in Figure 12. In this process, the low degree of freedom deformation serves as the initial value for the deformation in the next step. Therefore, local minima can be avoided more efficiently.

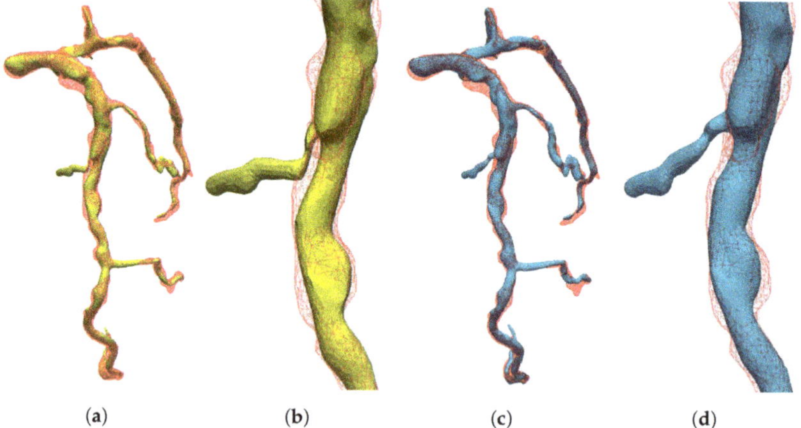

(a)   **(b)**   **(c)**   **(d)**

Figure 10. The effect of modified hyper-elastic regularization. We aligned the source model to the target model (red), which contains a loss of branch. The figures show effects: (**a**,**b**) without regularization (yellow) and (**c**,**d**) with regularization (blue).

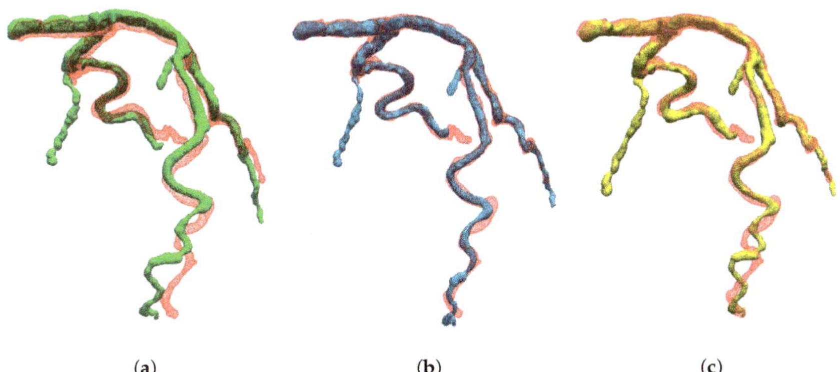

(**a**) (**b**) (**c**)

Figure 11. Qualitative comparison of GMM and the proposed method: (**a**) Initial template model (green) and target coronary artery (red); (**b**) result of the proposed method (blue); and (**c**) result of Gaussian mixture modeling with TPS+L_2 (yellow).

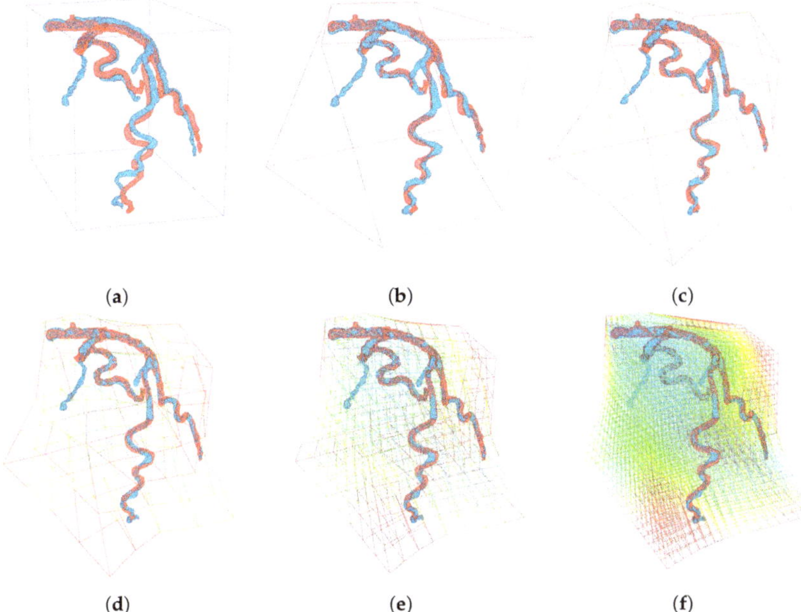

(**a**) (**b**) (**c**)

(**d**) (**e**) (**f**)

Figure 12. Qualitative comparison of the proposed method while changing the deformation resolution: (**a**) Initial template model (blue) and target model (red); (**b**–**f**) the results of registration (blue) at different cage resolutions from [1, 1, 1] to [5, 5, 5], respectively.

5.2.2. The Representation Power of Interpolated Model

The proposed shape interpolation method may have limited representation ability for intermediate shapes. This limited shape representation is due to the recurring shape of the adjacent phases. We observed that the shape interpolation method restrictively delineates the intermediate shape. The shapes of the neighboring phases to the target phase resemble each other, but the target shape and the neighbor shapes are considerably different, as shown in Figure 13e.

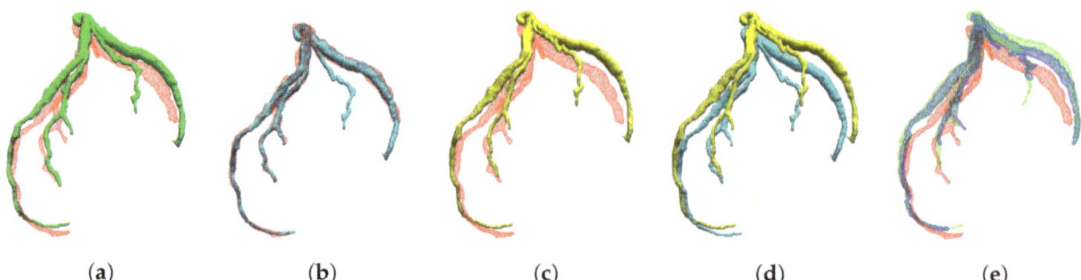

(a)   (b)   (c)   (d)   (e)

Figure 13. Comparison of the registered model and interpolated model for patient 8: (**a**) Template (green) and 95% coronary artery (red edges); (**b**) registered model (blue); (**c**) interpolated model (yellow); (**d**) comparison of registration interpolation; and (**e**) comparison of neighboring coronary artery models (90%, dark blue; 0%, light green).

6. Discussion and Conclusions

In this paper, we proposed a method for generating a 3D + time vessel model from 4D CT images that can be used in real-time. Our purpose was to create a 4D vascular model without mesh degeneration, interpolate the model at high speed, and express a more precise shape.

To create a 4D vascular model, we matched the diastolic coronary artery model with the coronary artery model in other phases through hierarchical cage deformation. During the registration process, hyper-elastic regularization was used as a shape preservation constraint. The shape control points obtained as a result of registration were interpolated into a cyclic cubic spline to create a 3D+t model. The shape change depends only on the control points of the cage. The rapid deformation application and the preservation features of the local information are beneficial in the shape registration process.

To evaluate the precision of the proposed method, quantitative and qualitative evaluation was performed on 160 CTA volumes acquired from eight patients. In the quantitative evaluation, we assessed:

1. The trade-off between the shape matching accuracy and calculation time according to the hierarchical deformation;
2. The comparative evaluation with other methods;
3. The accuracy of the shape interpolation model, according to the time sampling interval.

In the step of measuring the shape matching precision according to the hierarchical deformation, we observed that the matching precision converged in the fourth step of the hierarchical deformation, where the calculation time was 23.53 s on average. In the fifth deformation depth of hierarchical transformation, the matching accuracy slightly increased, but the required time increased by 28.70% (to 33.00 s). In the hierarchical transformation of the cage creation stage, the control points constituting the cage increased exponentially, as a regular grid was used. We obtained the mean distance with precision of a 0.543 mm and standard deviation 0.222 in step 4 of the hierarchical transformation, where the Dice coefficient obtained an average of 0.754 and a standard deviation of 0.064.

Compared with other algorithms, the GMM method requires the creation of a mixture model for each point and, so, even with the same degree of freedom of transformation, the calculation time was as high as 40 s for the GRBF model and 33 s for the TPS model. In addition, in the average distance index, TPS_L2 was 0.530 mm, which had an error lower than that (0.543) of the proposed method; however, when comparing the Dice coefficient, TPS_L2 was observed to be 0.690, 0.045 points lower than the index of the proposed method (0.735).

If the indicators of DC and AD values conflict with each other, it is necessary to determine which indicator is better to express the accuracy of the matching result; for example, (1) high DC value and high AD value or (2) low DC Value and low AD value. At this time, we decided case (1) was a better indicator. This was because, in the vascular model between 15% and 35%, which showed a great difference with the 75% phase, a difference in AD values between the algorithms was not significantly observed, but the difference in DC values was noticeable. This was not only consistently observed in the quantitative evaluation of patient data, but also explains the differences arising from inappropriate deformations occurring in excessively twisted blood vessels during the qualitative evaluation.

The precision of the shape interpolation model was measured with different phase-sampling sets, where the accuracy was worsened with a larger phase-sampling interval. However, this limitation may be resolved by increasing the temporal resolution and by co-operating with the other real-time imaging systems, such as X-ray angiography or 4D US.

A qualitative evaluation was performed to support the quantitative evaluation mentioned above, as well as to observe the effect of the proposed model on the visualization stage. In the qualitative evaluation, (1) the effect of the modified hyper-elastic regularization and hierarchical transformation, and (2) the limitations in shape interpolation were observed.

When we observed the effect of hyper-elastic regularization, the shape was transformed to be visually plausible when hyper-elastic regularization was used. This phenomenon was particularly well seen in the coronary artery model with loss of shape. This is a characteristic obtained by minimizing the excessive deformation by robustly obtaining face points and volume points against the rapid degeneration of the cage mesh due to incorrect correspondence pairs. Compared with other non-rigid algorithms, the proposed method was able to cope with the local minima that occur during the optimization process by hierarchically performing the transformation effectively. In particular, it was shown that, in the vascular model with an excessive twist, optimal deformation was gradually obtained from low-dimensional deformation.

In summary, the proposed 3D+t vascular modeling method utilized hierarchical deformation for robust shape registration, while interpolation of the registered vascular structure enabled the restoration of small and complex geometries due to the cardiac cycle.

The electric potential of the myocardium generates the ECG signal, and the electric potential of the myocardium is related to movement and contraction of the heart. As an observation tool of the myocardium, the proposed method can provide an alternative to real-time imaging by using the ECG signal and 4D CT. In intraoperative situations, invasive coronary angiography is a method for monitoring the movement of the coronary artery, which provides limited deformation information (up to the 2D plane). With the proposed method, we expect that 3D contraction and strain of the myocardium, according to the ECG signal, can be observed, as the coronary arteries are attached to the epicardium. The ability of motion monitoring is directly related to the evaluation of the physiological function of the myocardium.

In addition, the modified hyper-elastic regularization prevented implausible deformation and mesh degeneration, which must be avoided in the analysis of 4D blood flow. We demonstrated the accuracy of the proposed method by presenting qualitative and quantitative evaluations using data from eight patients.

Therefore, we expect that the proposed 3D+t vascular model can be utilized in real-time applications such as (1) pre-operative blood flow analysis, which requires rapid shape creation without mesh degeneration; and (2) 2D invasive coronary angiography-3D shape registration during the intervention, where it can be used as an image guidance tool that provides real-time shape information according to the ECG signal during percutaneous coronary intervention.

As a limitation of this study, since the deformed model of the proposed model is limited to a 3D mesh, blood vessel segmentation is required in phases other than the template model. In a future study, we will conduct a study on a volume–template mesh model matching method which can be applied to volumetric data, including shape loss, in order to eliminate unnecessary repetitive processes.

Author Contributions: Conceptualization, S.Y., C.Y., E.J.C. and D.L.; Data curation, S.Y., C.Y. and E.J.C.; Formal analysis, S.Y. and D.L.; Funding acquisition, D.L.; Investigation, S.Y.; Methodology, S.Y. and D.L.; Project administration, D.L.; Resources, S.Y., C.Y., E.J.C. and D.L.; Software, S.Y.; Supervision, D.L.; Validation, S.Y.; Visualization, S.Y.; Writing—original draft, S.Y.; Writing—review & editing, S.Y. and D.L. All authors have read and agreed to the published version of the manuscript.

Funding: This work was supported by KIST intramural grants (2E30260) and the Ministry of Trade Industry & Energy (MOTIE, Korea), Ministry of Science & ICT (MSIT, Korea), and Ministry of Health & Welfare (MOHW, Korea) under Technology Development Program for AI-Bio-Robot-Medicine Convergence (20001655).

Conflicts of Interest: The authors declare no conflict of interest.

Appendix A. Patient-Wise Evaluations

Appendix A.1. Dice Coefficients at Different Levels of Deformation

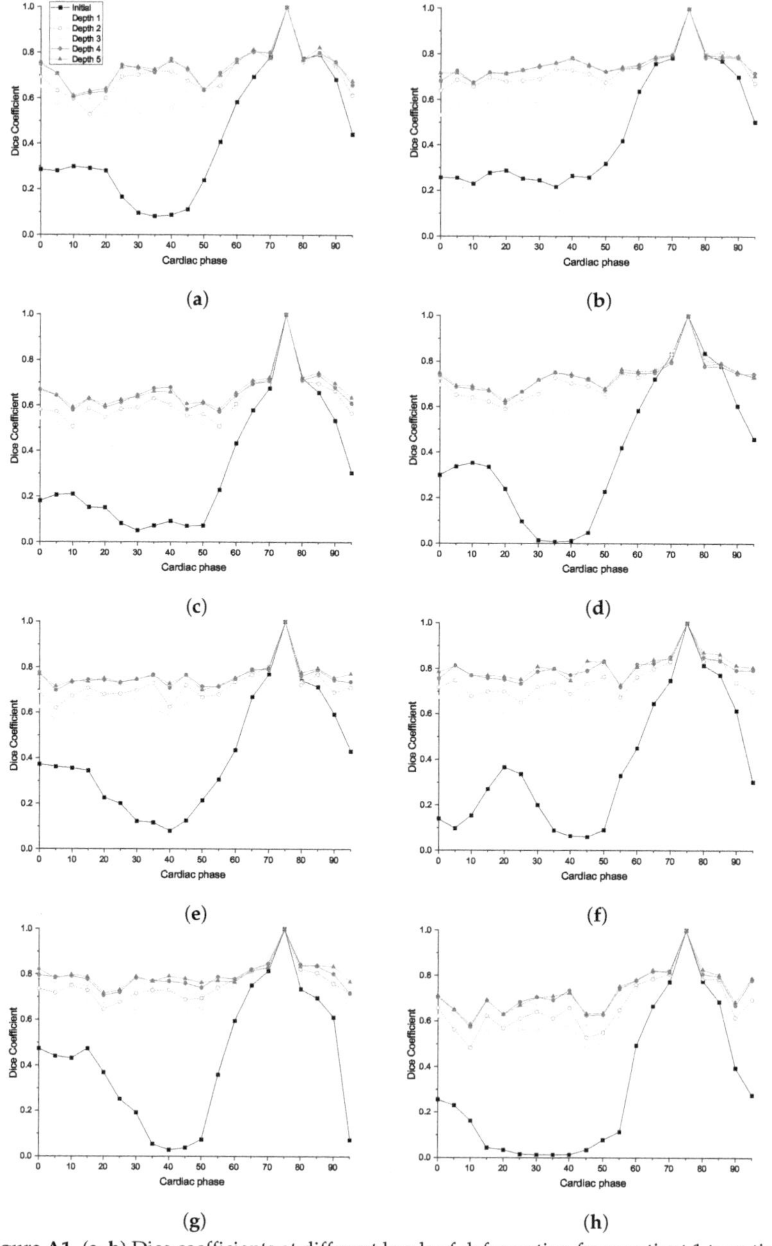

Figure A1. (a–h) Dice coefficients at different levels of deformation from patient 1 to patient 8.

Appendix A.2. Average Distances at Different Levels of Deformation

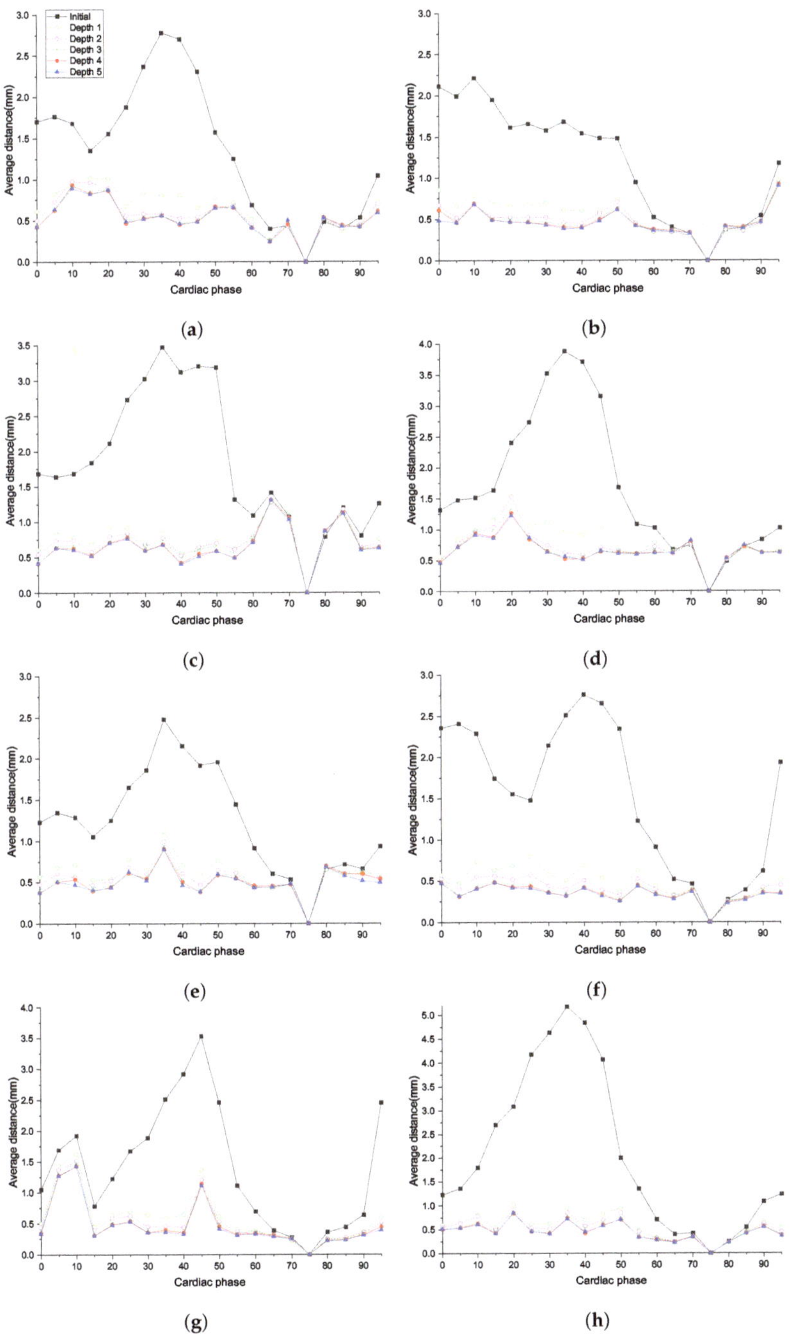

Figure A2. (**a–h**) Average distances between target model and deformed models for Patients 1–8, respectively.

Appendix A.3. Dice Coefficients for Different Methods

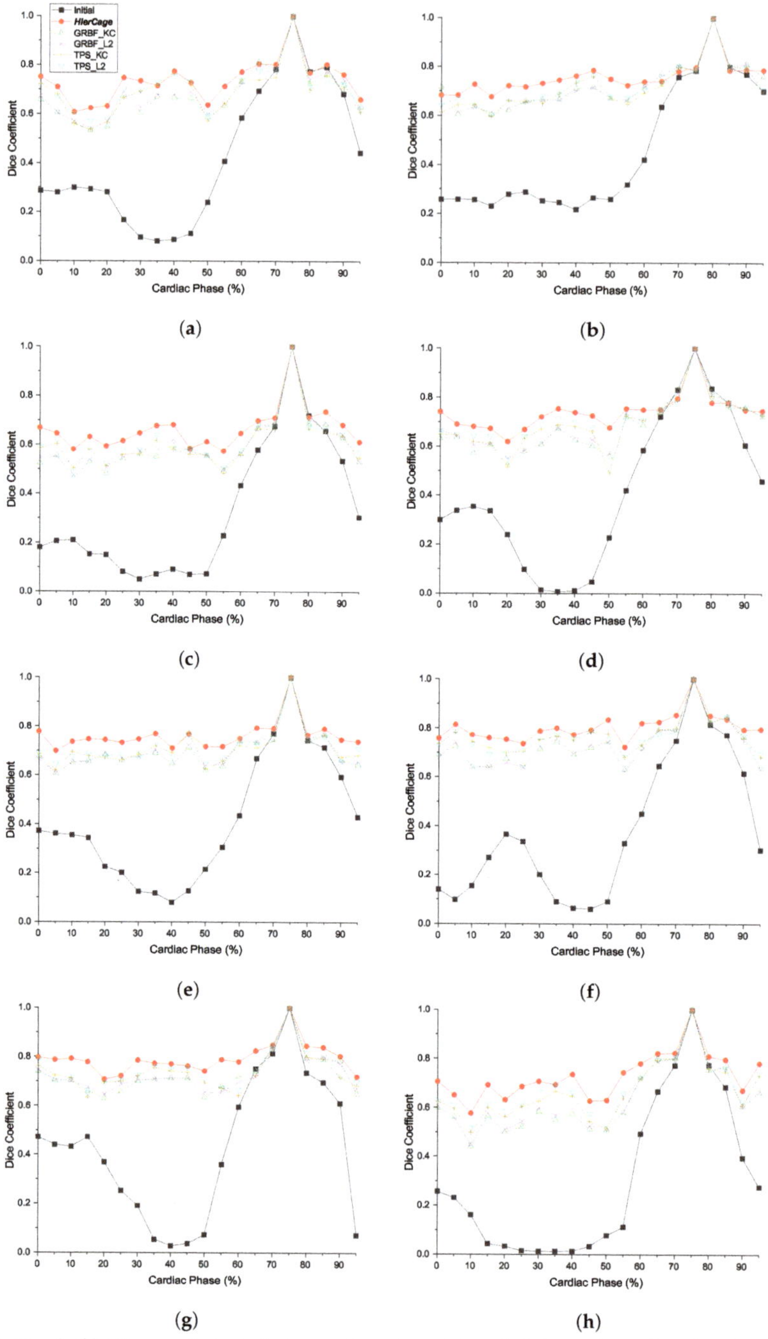

Figure A3. (**a–h**) Dice coefficients between target model and deformed models for the different cardiac phases.

Appendix A.4. Average Distances for Different Methods

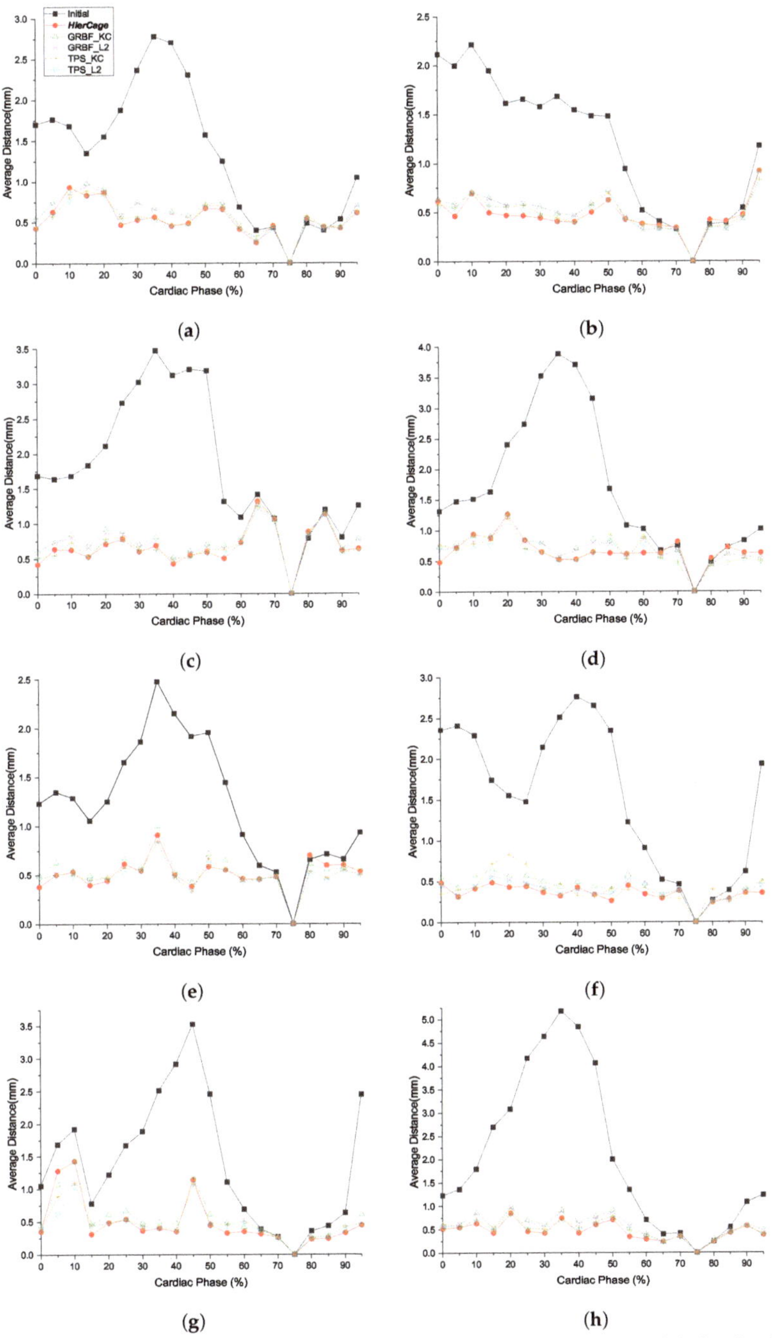

Figure A4. (**a–h**) Average distances between target model and deformed models for the different cardiac phases.

Appendix A.5. Dice Coefficients for Different Phase Sampling Methods

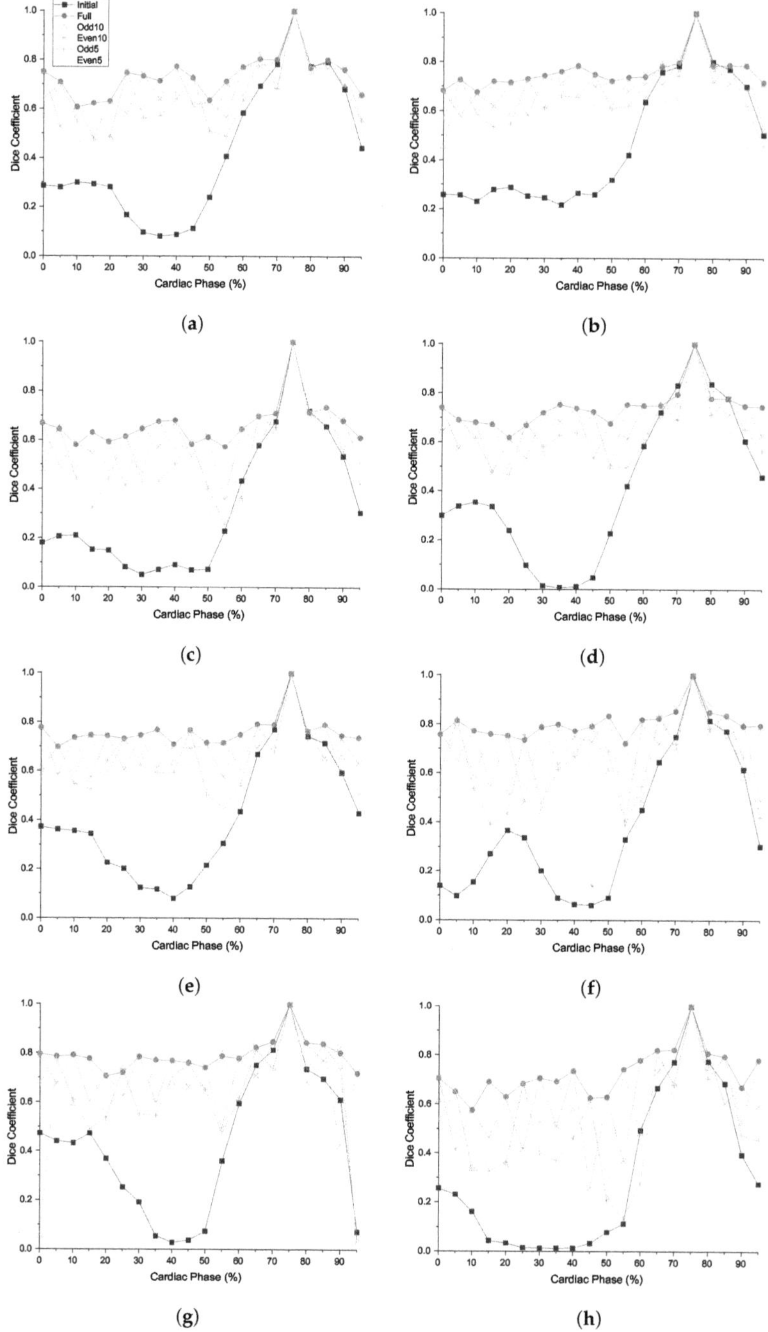

Figure A5. (**a–h**) Dice coefficients between target model and deformed models for the different cardiac phases.

Appendix A.6. Average Distances for Different Phase Sampling Methods

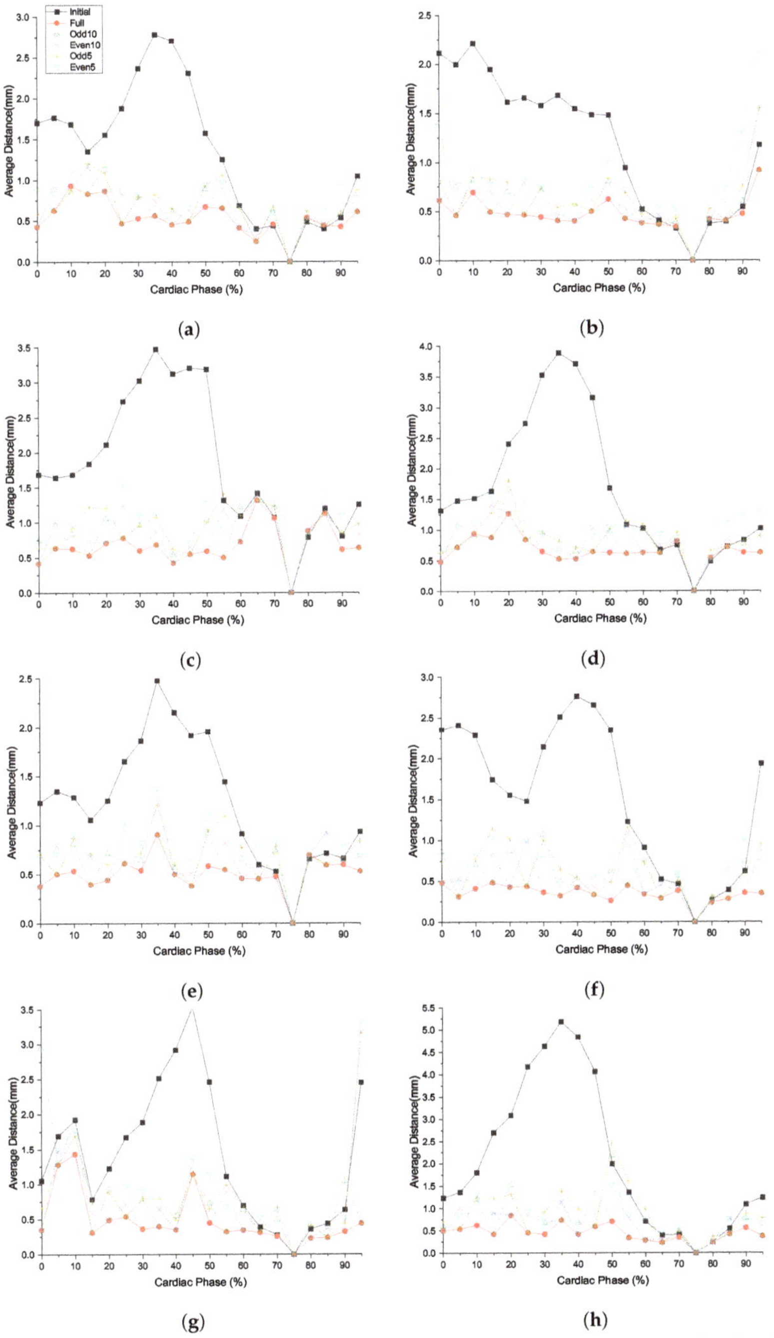

Figure A6. (**a–h**) Average distances between target model and deformed models for the different cardiac phases.

References

1. Virani, S.S.; Alonso, A.; Benjamin, E.J.; Bittencourt, M.S.; Callaway, C.W.; Carson, A.P.; Chamberlain, A.M.; Chang, A.R.; Cheng, S.; Delling, F.N.; et al. Heart disease and stroke statistics—2020 update a report from the American Heart Association. *Circulation* **2020**, E139–E596. [CrossRef] [PubMed]
2. Hadjiiski, L.; Zhou, C.; Chan, H.P.; Chughtai, A.; Agarwal, P.; Kuriakose, J.; Kazerooni, E.; Wei, J.; Patel, S. Coronary CT angiography (cCTA): Automated registration of coronary arterial trees from multiple phases. *Phys. Med. Biol.* **2014**, *59*, 4661. [CrossRef] [PubMed]
3. Zeng, S.; Feng, J.; An, Y.; Lu, B.; Lu, J.; Zhou, J. Towards Accurate and Complete Registration of Coronary Arteries in CTA Images. In Proceedings of the International Conference on Medical Image Computing and Computer-Assisted Intervention, Granada, Spain, 16–20 September 2018; pp. 419–427.
4. Biglarian, M.; Larimi, M.M.; Afrouzi, H.H.; Moshfegh, A.; Toghraie, D.; Javadzadegan, A.; Rostami, S. Computational investigation of stenosis in curvature of coronary artery within both dynamic and static models. *Comput. Meth. Programs Biomed.* **2020**, *185*, 105170. [CrossRef] [PubMed]
5. Wu, X.; von Birgelen, C.; Muramatsu, T.; Li, Y.; Holm, N.R.; Reiber, J.H.; Tu, S. A novel four-dimensional angiographic approach to assess dynamic superficial wall stress of coronary arteries in vivo: initial experience in evaluating vessel sites with subsequent plaque rupture. *EuroIntervention* **2017**, *13*, e1099–e1103. [CrossRef]
6. Elattar, M.A.; Vink, L.W.; van Mourik, M.S.; Baan Jr, J.; vanBavel, E.T.; Planken, R.N.; Marquering, H.A. Dynamics of the aortic annulus in 4D CT angiography for transcatheter aortic valve implantation patients. *PLoS ONE* **2017**, *12*, e0184133. [CrossRef]
7. Shi, B.; Katsevich, G.; Chiang, B.S.; Katsevich, A.; Zamyatin, A. Image registration for motion estimation in cardiac CT. In Proceedings of the 2014 SPIE Medical Imaging, San Diego, CA, USA, 15–20 February 2014; Volume 9033, p. 90332E.
8. Forte, M.N.V.; Valverde, I.; Prabhu, N.; Correia, T.; Narayan, S.A.; Bell, A.; Mathur, S.; Razavi, R.; Hussain, T.; Pushparajah, K.; et al. Visualization of coronary arteries in paediatric patients using whole-heart coronary magnetic resonance angiography: comparison of image-navigation and the standard approach for respiratory motion compensation. *J. Cardiovasc. Magn. Reson.* **2019**, *21*, 1–9.
9. Coppo, S.; Piccini, D.; Bonanno, G.; Chaptinel, J.; Vincenti, G.; Feliciano, H.; Van Heeswijk, R.B.; Schwitter, J.; Stuber, M. Free-running 4D whole-heart self-navigated golden angle MRI: Initial results. *Magn. Reson. Med.* **2015**, *74*, 1306–1316. [CrossRef]
10. Li, S.; Xie, Z.; Xia, Q.; Hao, A.; Qin, H. Hybrid 4D cardiovascular modeling based on patient-specific clinical images for real-time PCI surgery simulation. *Graph. Model.* **2019**, *101*, 1–7. [CrossRef]
11. Lamash, Y.; Fischer, A.; Carasso, S.; Lessick, J. Strain analysis from 4-D cardiac CT image data. *IEEE Trans. Biomed. Eng.* **2014**, *62*, 511–521. [CrossRef]
12. Gupta, V.; Lantz, J.; Henriksson, L.; Engvall, J.; Karlsson, M.; Persson, A.; Ebbers, T. Automated three-dimensional tracking of the left ventricular myocardium in time-resolved and dose-modulated cardiac CT images using deformable image registration. *J. Cardiovasc. Comput. Tomogr.* **2018**, *12*, 139–148.
13. Li, Q.; Tong, Y.; Yin, Y.; Cheng, P.; Gong, G. Definition of the margin of major coronary artery bifurcations during radiotherapy with electrocardiograph-gated 4D-CT. *Phys. Med.* **2018**, *49*, 90–94. [CrossRef] [PubMed]
14. Liu, B.; Bai, X.; Zhou, F. Local motion-compensated method for high-quality 3D coronary artery reconstruction. *Biomed. Opt. Express* **2016**, *7*, 5268–5283. [CrossRef] [PubMed]
15. Chen, M.Y.; Shanbhag, S.M.; Arai, A.E. Submillisievert median radiation dose for coronary angiography with a second-generation 320–detector row CT scanner in 107 consecutive patients. *Radiology* **2013**, *267*, 76–85. [CrossRef] [PubMed]
16. Besl, P.J.; McKay, N.D. Method for registration of 3-D shapes. In Proceedings of the Robotics '91, Boston, MA, USA, 14–15 November 1991.
17. Sorkine, O.; Alexa, M. As-rigid-as-possible surface modeling. In Proceedings of the Symposium on Geometry Processing, Barcelona, Spain, 4–6 July 2007; Volume 4, pp. 109–116.
18. Davatzikos, C. Spatial transformation and registration of brain images using elastically deformable models. *Comput. Vis. Image Underst.* **1997**, *66*, 207–222. [CrossRef]

19. Pennec, X.; Stefanescu, R.; Arsigny, V.; Fillard, P.; Ayache, N. Riemannian elasticity: A statistical regularization framework for non-linear registration. In Proceedings of the International Conference on Medical Image Computing and Computer-Assisted Intervention, Palm Springs, CA, USA, 26–29 October 2005; pp. 943–950.

20. Burger, M.; Modersitzki, J.; Ruthotto, L. A hyperelastic regularization energy for image registration. *SIAM J. Sci. Comput.* **2013**, *35*, B132–B148. [CrossRef]

21. Chiang, M.C.; Leow, A.D.; Klunder, A.D.; Dutton, R.A.; Barysheva, M.; Rose, S.E.; McMahon, K.L.; De Zubicaray, G.I.; Toga, A.W.; Thompson, P.M. Fluid registration of diffusion tensor images using information theory. *IEEE Trans. Med. Imaging* **2008**, *27*, 442–456. [CrossRef]

22. Vercauteren, T.; Pennec, X.; Perchant, A.; Ayache, N. Symmetric log-domain diffeomorphic registration: A demons-based approach. In Proceedings of the International Conference on Medical Image Computing and Computer-Assisted Intervention, New York, NY, USA, 6–10 September 2008; pp. 754–761.

23. Yeo, B.T.; Vercauteren, T.; Fillard, P.; Peyrat, J.M.; Pennec, X.; Golland, P.; Ayache, N.; Clatz, O. DT-REFinD: Diffusion tensor registration with exact finite-strain differential. *IEEE Trans. Med. Imaging* **2009**, *28*, 1914–1928. [CrossRef]

24. Younes, L.; Qiu, A.; Winslow, R.L.; Miller, M.I. Transport of relational structures in groups of diffeomorphisms. *J. Math. Imaging Vis.* **2008**, *32*, 41–56. [CrossRef]

25. Cootes, T.F.; Taylor, C.J.; Cooper, D.H.; Graham, J. Active shape models-their training and application. *Comput. Vis. Image Underst.* **1995**, *61*, 38–59. [CrossRef]

26. Glocker, B.; Komodakis, N.; Navab, N.; Tziritas, G.; Paragios, N. Dense registration with deformation priors. In Proceedings of the International Conference on Information Processing in Medical Imaging, Williamsburg, VA, USA, 5–10 July 2009; pp. 540–551.

27. Baka, N.; Metz, C.; Schultz, C.; Neefjes, L.; van Geuns, R.J.; Lelieveldt, B.P.; Niessen, W.J.; van Walsum, T.; de Bruijne, M. Statistical coronary motion models for 2D+ t/3D registration of X-ray coronary angiography and CTA. *Med. Image Anal.* **2013**, *17*, 698–709. [CrossRef]

28. Yang, X.; Xue, Z.; Liu, X.; Xiong, D. Topology preservation evaluation of compact-support radial basis functions for image registration. *Pattern Recognit. Lett.* **2011**, *32*, 1162–1177. [CrossRef]

29. Donato, G.; Belongie, S. Approximate thin plate spline mappings. In Proceedings of the European Conference on Computer Vision, Copenhagen, Denmark, 28–31 May 2002; pp. 21–31.

30. Sederberg, T.W.; Parry, S.R. Free-form deformation of solid geometric models. In Proceedings of the 13th Annual Conference on Computer Graphics and Interactive Techniques, Dallas, TX, USA, August 18–22 1986; pp. 151–160.

31. Sdika, M. A fast nonrigid image registration with constraints on the Jacobian using large scale constrained optimization. *IEEE Trans. Med. Imaging* **2008**, *27*, 271–281. [CrossRef] [PubMed]

32. Rueckert, D.; Aljabar, P.; Heckemann, R.A.; Hajnal, J.V.; Hammers, A. Diffeomorphic registration using B-splines. In Proceedings of the International Conference on Medical Image Computing and Computer-Assisted Intervention, Copenhagen, Denmark, 1–6 October 2006; pp. 702–709.

33. Chui, H.; Rangarajan, A. A new point matching algorithm for non-rigid registration. *Comput. Vis. Image Underst.* **2003**, *89*, 114–141. [CrossRef]

34. Chui, H.; Rangarajan, A. A feature registration framework using mixture models. In Proceedings of the IEEE Workshop on Mathematical Methods in Biomedical Image Analysis (MMBIA-2000), Hilton Head Island, SC, USA, 11–12 June 2000; pp. 190–197.

35. Jian, B.; Vemuri, B.C. A robust algorithm for point set registration using mixture of Gaussians. In Proceedings of the 10th IEEE International Conference on Computer Vision (ICCV'05), Las Vegas, NV, USA, 17–21 October 2005; pp. 1246–1251.

36. Myronenko, A.; Song, X.; Carreira-Perpinán, M.A. Non-rigid point set registration: Coherent point drift (CPD). *Adv. Neural Inf. Process. Syst.* **2007**, *1*, 1009–1016.

37. Myronenko, A.; Song, X. Point set registration: Coherent point drift. *IEEE Trans. Pattern Anal. Mach. Intell.* **2010**, *32*, 2262–2275. [CrossRef] [PubMed]

38. Jian, B.; Vemuri, B.C. Robust point set registration using gaussian mixture models. *IEEE Trans. Pattern Anal. Mach. Intell.* **2010**, *33*, 1633–1645. [CrossRef]

39. Ma, J.; Qiu, W.; Zhao, J.; Ma, Y.; Yuille, A.L.; Tu, Z. Robust L_2E estimation of transformation for non-rigid registration. *IEEE Trans. Signal Process.* **2015**, *63*, 1115–1129. [CrossRef]

40. Ma, J.; Zhao, J.; Yuille, A.L. Non-rigid point set registration by preserving global and local structures. *IEEE Trans. Image Process.* **2015**, *25*, 53–64.

41. Yushkevich, P.A.; Piven, J.; Cody Hazlett, H.; Gimpel Smith, R.; Ho, S.; Gee, J.C.; Gerig, G. User-Guided 3D Active Contour Segmentation of Anatomical Structures: Significantly Improved Efficiency and Reliability. *Neuroimage* **2006**, *31*, 1116–1128. [CrossRef]

42. Seifarth, H.; Wienbeck, S.; Pusken, M.; Juergens, K.U.; Maintz, D.; Vahlhaus, C.; Heindel, W.; Fischbach, R. Optimal systolic and diastolic reconstruction windows for coronary CT angiography using dual-source CT. *Am. J. Roentgenol.* **2007**, *189*, 1317–1323. [CrossRef]

43. Press, W.H.; Teukolsky, S.A.; Vetterling, W.T.; Flannery, B.P. *Numerical Recipes 3rd Edition: The Art of Scientific Computing*; Cambridge University Press: New York, NY, USA, 2007.

44. Dagum, L.; Ramesh, M. OpenMP: an industry standard API for shared-memory programming. *IEEE Comput. Sci. Eng.* **1998**, *5*, 46–55. [CrossRef]

45. Pheatt, C. Intel threading building blocks. *J. Comput. Sci. Coll.* **2008**, *23*, 298–298.

Review

Wearable Sensors Incorporating Compensatory Reserve Measurement for Advancing Physiological Monitoring in Critically Injured Trauma Patients

Victor A. Convertino [1,2,*], Steven G. Schauer [1,2,3], Erik K. Weitzel [2,3,4] , Sylvain Cardin [5], Mark E. Stackle [6], Michael J. Talley [7], Michael N. Sawka [8] and Omer T. Inan [8]

[1] Battlefield Health & Trauma Center for Human Integrative Physiology, US Army Institute of Surgical Research, JBSA Fort Sam Houston, San Antonio, TX 78234, USA; steven.g.schauer.mil@mail.mil

[2] Uniformed Services University of the Health Sciences, Bethesda, MD 20814, USA; erik.k.weitzel.mil@mail.mil

[3] Brooke Army Medical Center, JBSA Fort Sam Houston, San Antonio, TX 78234, USA

[4] 59th Medical Wing, JBSA Lackland, San Antonio, TX 78236, USA

[5] Navy Medical Research Unit, San Antonio, TX 78234, USA; sylvain.cardin.civ@mail.mil

[6] US Army Institute of Surgical Research, JBSA Fort Sam Houston, San Antonio, TX 78234, USA; mark.e.stackle.mil@mail.mil

[7] US Army Medical Research and Development Command, Fort Detrick, Frederick, MD 21702, USA; michael.j.talley4.mil@mail.mil

[8] Georgia Institute of Technology, Atlanta, GA 30332, USA; michael.sawka@biosci.gatech.edu (M.N.S.); inan@gatech.edu (O.T.I.)

* Correspondence: victor.a.convertino.civ@mail.mil

Received: 29 September 2020; Accepted: 4 November 2020; Published: 10 November 2020

Abstract: Vital signs historically served as the primary method to triage patients and resources for trauma and emergency care, but have failed to provide clinically-meaningful predictive information about patient clinical status. In this review, a framework is presented that focuses on potential wearable sensor technologies that can harness necessary electronic physiological signal integration with a current state-of-the-art predictive machine-learning algorithm that provides early clinical assessment of hypovolemia status to impact patient outcome. The ability to study the physiology of hemorrhage using a human model of progressive central hypovolemia led to the development of a novel machine-learning algorithm known as the compensatory reserve measurement (CRM). Greater sensitivity, specificity, and diagnostic accuracy to detect hemorrhage and onset of decompensated shock has been demonstrated by the CRM when compared to all standard vital signs and hemodynamic variables. The development of CRM revealed that continuous measurements of changes in arterial waveform features represented the most integrated signal of physiological compensation for conditions of reduced systemic oxygen delivery. In this review, detailed analysis of sensor technologies that include photoplethysmography, tonometry, ultrasound-based blood pressure, and cardiogenic vibration are identified as potential candidates for harnessing arterial waveform analog features required for real-time calculation of CRM. The integration of wearable sensors with the CRM algorithm provides a potentially powerful medical monitoring advancement to save civilian and military lives in emergency medical settings.

Keywords: wearable sensors; physiology; medical monitoring; vital signs; compensatory reserve

1. Introduction

Vital signs are the most rudimentary, yet frequently relied upon physiologic data used by emergency care clinicians on which they base treatment decisions. In both prehospital and emergency department settings, vital signs are used as a primary method for triaging patients and resources for both trauma and medical encounters [1]. Pulse palpation and blood pressure have been used by physicians dating back to the 18th century with the documented work of Stephen Hales [2]. Whereas anatomical imaging diagnostics have enjoyed major advancement with novel diagnostic modalities such as computed tomography and magnetic resonance imaging in the hospital, physiological monitoring available in the prehospital and emergency room settings has remained largely unchanged. Blood pressure is still measured with a sphygmomanometer with only small incremental gains in technology over the last century. In austere clinical settings where sphygmomanometry may not be readily available (e.g., military operations, wilderness medicine), patient status is assessed by gross manual measures such as palpitation for radial pulse character and mental status [3–6]. In this regard, the Special Operations Medical Association Prolonged Field Care Working Group identified a "monitor to provide hands-free vital signs data at regular intervals" as a core capability needed to meet the requirement for prolonged field care on the battlefield [7,8]. New advancements in capturing and analyzing real-time electronic signals from the body using wearable sensor signals that are integrated with advanced computer processing capabilities hold great promise for development of novel monitoring technologies. In this review, we provide evidence for the need to use a photoplethysmographic (PPG) signal as the most informative 'vital sign' to be captured in emergency medical care settings. We introduce a variety of currently available wearable sensor technologies that could be used to harness PPG signals for integration with a novel predictive machine-learning algorithm designed to optimize pathophysiological monitoring and early triage decision support beyond standard vital signs.

2. Need to Identify New Vital Sign Measurements

2.1. Compromised DO_2—A Primary Clinical Challenge to Effective Medical Monitoring

Hemorrhage is the primary reason for death after major trauma in both civilian and military settings [9–13]. If not controlled in its early stages, hemorrhage can result in inadequate systemic oxygen delivery (DO_2) to vital organs (e.g., brain, heart, gut) that without effective intervention can rapidly lead to organ dysfunction and tissue death [14]. Clinically, DO_2 is indirectly assessed by measurements of standard vital signs such as blood pressure. However, improvement in blood pressure alone does not correlate with oxygen received at the tissue level as supported by the observation that crystalloid fluids can elevate systolic pressure while simultaneously worsening patient outcomes [9]. Clinicians frequently define a measurement of systolic pressure <90 mmHg as hypotension incapable of sustaining adequate DO_2. More recent data suggest the use of this threshold may not accurately represent risk for poor clinical outcome [4,15,16]. To this end, it has been proposed that optimal assessment of patient status demands actual measurement of systemic DO_2 [17–19].

2.2. Current Vital Sign Monitoring

One limitation of most modern monitoring systems is a bias toward capture of only standard vital signs (Table 1). Standard vital signs exhibit little change during the early stages of volume loss due to physiological compensatory responses [20–25]. Such responses (e.g., deep inspiration, tachycardia, and vasoconstriction) regulate and maintain blood pressure and tissue perfusion prior to the onset of decompensated shock during the early stages of hemorrhage, sepsis, dehydration, and other forms of central hypovolemia [26–29].

Table 1. Qualitative timing of changes in traditional vital signs and blood chemistries during progressive central hypovolemia. Modified from Convertino et al. [14,22] and Moulton et al. [30].

Vital Sign or Measurement	Change During Progressive Central Hypovolemia
Systolic blood pressure	Late
Diastolic blood pressure	Late
Mean blood pressure	Late
Heart rate	Non-specific
Shock index (heart rate/systolic pressure)	Late
Oxygen saturation	Late
Radial pulse character assessment	Late
End-tidal CO_2	Late, Non-specific
Respiratory rate	Late, Non-specific
Glasgow Coma Scale	Late, Non-specific
Blood pH, PCO_2, Base Excess	Late, Non-specific
Blood Lactate	Late, Non-specific
Hematocrit, Hemoglobin	Late, Non-specific

In an effort to identify and compare the time course of changes in standard vital signs and physiological compensatory responses during the early stages of blood loss, lower body negative pressure (LBNP) has emerged as a validated model for controlled progressive reductions in central blood volume that mimics the physiology of hemorrhage in humans [14,31,32]. Like hemorrhage, LBNP leads to reduced filling of the heart which in turn reduces cardiac stroke volume and output, resulting in lower DO_2 (Figure 1). Using this model of human hemorrhage has consistently revealed that commonly relied upon vital signs are not specific to the condition of blood loss or do not change until too late in the clinical course of reduced central blood volume to allow optimized patient care (Table 1).

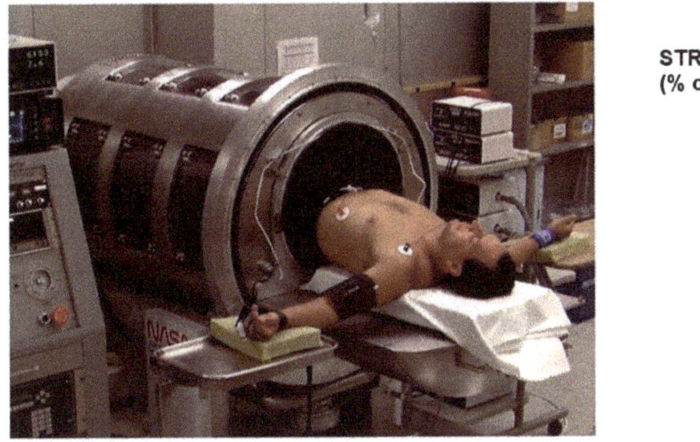
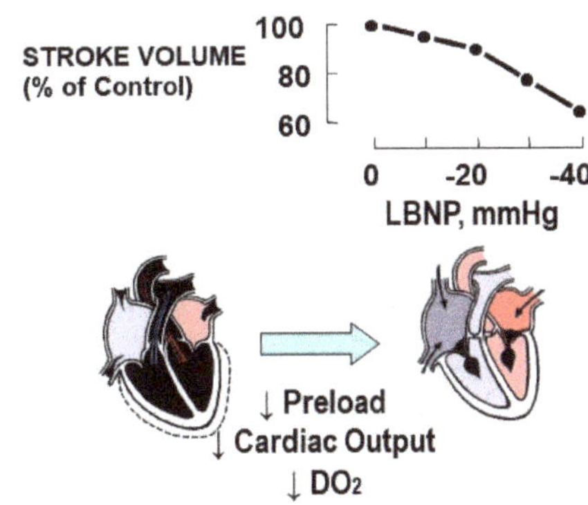

Figure 1. Human subject placed in the lower body negative pressure (LBNP) chamber used to induce progressive reductions in cardiac filling (preload), stroke volume, cardiac output and DO_2 similar to hemorrhage.

Currently used monitors track limited vital sign measurements and chemistries on an interval basis (e.g., blood pressure) with limited capability for providing continuous, real-time physiological assessments (e.g., electrocardiogram, pulse, oxygen saturation) and are often non-specific to magnitude of hypovolemia. In addition, current commercial monitoring systems for emergency settings are bulky, power hungry, have wires interfering with patient care, and sensitive to motion artifact. Despite the above findings and limitations, clinicians continue to rely upon standard vital signs or blood chemistries when deciding to intervene because new and more effective monitoring technologies are not available.

2.3. Accuracy, Sensitivity, and Specificity

The effectiveness of any monitoring technology relies on its ability to provide accurate, sensitive, and specific information about the clinical condition of the patient. In this regard, it is critical that there be an assessment of the number of cases correctly identified as unhealthy (True Positive or *TP* rate), correctly identified as healthy (True Negative or *TN* rate), incorrectly identified as healthy (False Negative or *FN*), and incorrectly identified as unhealthy (False Positive or *FP*). Of course, all of this requires an agreed upon reference gold standard. Once these parameters are quantified, accuracy, sensitivity and specificity of the measurement can be assessed. Within this framework, an estimation of accuracy can be calculated as the ratio of true positive and negative cases (*TP* plus *TN*) to the sum of all measured cases. Mathematically, this can be stated as:

$$\text{Accuracy} = \frac{TP + TN}{TP + TN + FP + FN} \tag{1}$$

Since sensitivity of a measurement represents its ability to correctly identify unhealthy cases, it can be calculated as the ratio of *TP* to the sum of both true positive and false negative unhealthy cases. Mathematically, sensitivity can be stated as:

$$\text{Sensitivity} = \frac{TP}{TP + FN} \tag{2}$$

Specificity refers to the ability of a diagnostic modality to correctly identify or predict those individuals who are healthy. That is, specificity can be calculated as the ratio of *TN* to the sum of all healthy cases and can be stated mathematically as:

$$\text{Specificity} = \frac{TN}{TN + FP} \tag{3}$$

In addition to accuracy, sensitivity and specificity, Youden's J statistic was first described in 1950 as a way to capture a single measurement performance assessment of a dichotomous diagnostic test [33]. The Youden's J statistic is calculated as:

$$J = \frac{TP}{TP + FN} + \frac{TN}{TN + FP} - 1 \tag{4}$$

Or in its simplistic form:

$$J = Sensitivity + Specificity - 1 \tag{5}$$

The values of the J statistic range from 0 to 1. A test that has as zero value gives the same proportion of positive results for both those with the disease state and those without the disease state. In other words, a J value of 0 is useless in assessing the status of a patient because it provides a positive result for the same number of patients that are experiencing the disease state as those that are not. Conversely, a J value of 1 demonstrates that an assessment modality accurately identifies all subjects with a disease state and those without. Quantitative comparisons of sensitivity, specificity, and the J statistic will be used in a subsequent section of this review for comparisons between standard vital signs and hemodynamic measurements in order to identify those physiological signals required by wearable sensors to optimize diagnostic accuracy.

3. New Monitoring Approach: The Compensatory Reserve

3.1. Defining the Compensatory Reserve

The compensatory reserve is a concept that represents the sum total of all physiological mechanisms that contribute to the maintenance of systemic DO_2 to the body's tissue. Conceptually, a compensatory reserve can be calculated as the difference between a baseline value at rest (100% reserve) and the value at the onset of hemodynamic instability (i.e., 0% reserve) [14,22–25,30,34–36]. In this regard, each individual has a finite 'reserve' consisting of physiological feedback mechanisms designed to compensate for low blood flow states. The complexity of this compensatory reserve is reflected by the reported observation that the physiology of integrated compensation is unique for each individual [37]. When this capacity to compensate becomes depleted, a state of decompensated shock occurs. Clinically, a compensatory reserve measurement (CRM) can be obtained from assessment of changing arterial pressure waveform morphology associated with changes in compensation [14,18,21,23,30,38–54]. Figure 2 illustrates that each arterial waveform consists of two primary waves: (1) an 'ejected' wave with features that are dictated by all compensatory mechanisms that influence myocardial function; and (2) a 'reflective' wave with features that are influenced by all compensatory mechanisms involved in the control of peripheral blood flow [14,22]. The LBNP model of hemorrhage has been used to generate a large reference database of more than 650,000 arterial pressure analog waveforms generated from noninvasive photoplethysmographic techniques and collected from more than 260 healthy men and women with age range of 18 to 55 years across various stages of reduced central blood volume to the point of decompensated shock (0% compensatory reserve) [22]. With application of advanced machine-learning technology to this large physiologically-diverse database, the CRM algorithm has 'learned' to provide rapid and continuous measures of changing arterial pressure waveform morphology to the clinical caregiver with the ability to gain an early and accurate assessment of the individual patient's medical status without the need for demographic data or measures of the patient's baseline physiology (as depicted in Figure 3) [21,24].

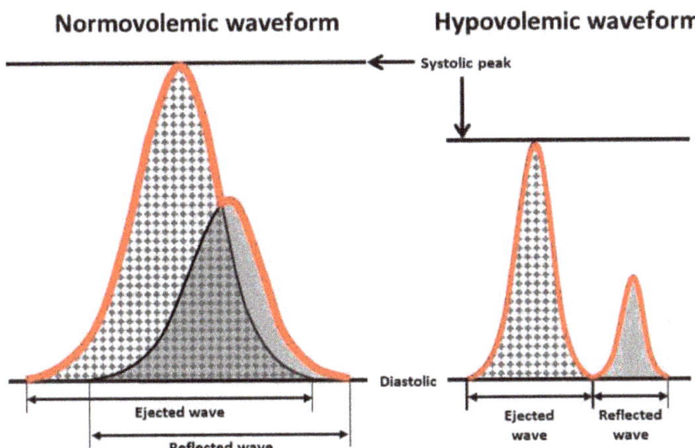

Figure 2. Illustration of changes in features of the ejected and reflected arterial waveforms progressing from a normal blood volume state to a state of reduced central blood volume (i.e., hypovolemia) such as that experienced during hemorrhage. The red line indicates the integrated waveform that is clinically observed. Modified from Convertino et al. [14,22,23].

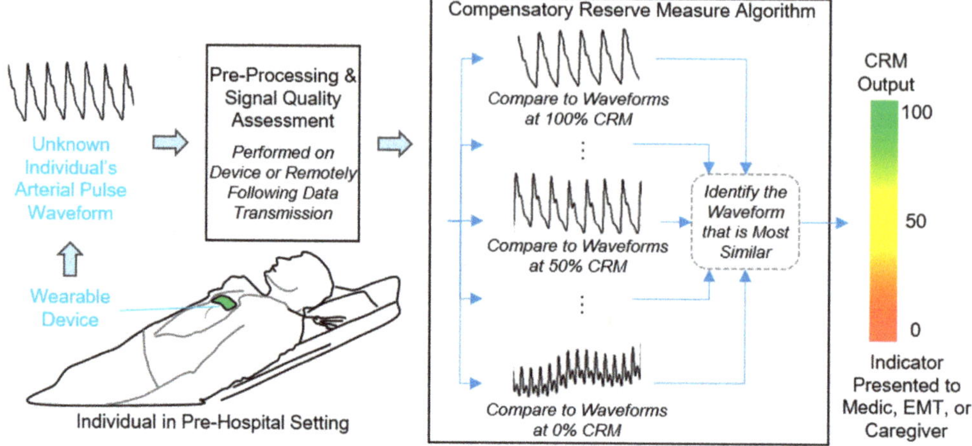

Figure 3. Diagram illustrating the overall framework envisioned for using the compensatory reserve measurement (CRM) in pre-hospital care, including the details on the CRM machine learning algorithm for assessing beat-to-beat analog arterial pressure waveform features in an individual patient unknown to the algorithm. The unknown arterial waveform is compared to a large waveform "library" collected from a diversity of human subjects exposed to progressive reductions in central blood volume. The algorithm identifies the most similar waveform in the waveform library with the unknown sample to generate a CRM value. Modified from Convertino et al. [14,18,21–23,55].

3.2. Performance Comparisons: Compensatory Reserve versus Vital Signs

Clinical measurements that inform and change the medical management of critically injured and sick patients should demonstrate high diagnostic accuracy. One approach to assess diagnostic accuracy includes direct comparisons of sensitivity and specificity across various monitoring capabilities. Table 2 presents such comparisons for the prediction power of standard vital signs and hemodynamic measurements for the onset of decompensated shock from data generated from LBNP experiments. The measurement of compensatory reserve displayed by far the greatest sensitivity, indicating its superior ability to correctly predict the onset of decompensated shock. Similarly, a greater specificity generated from the measurement of compensatory reserve indicated its superiority compared to the other vital signs and hemodynamic measures in the ability to identify patients who will not experience decompensated shock. The higher specificity of CRM reflects the failure of standard vital signs and hemodynamic measures alone to recognize the difference between individuals who are 'good' compensators from those who are 'poor' compensators [18,21–23,31,56–59]. Perhaps most striking is that standard vital signs and hemodynamic measurements have consistently been shown to lack sufficient accuracy as diagnostic tools to provide reliable clinical information [18,23,38,39,54,60,61]. In contrast, the ability of CRM to provide early reliable information with acceptable diagnostic accuracy is reflected by it being the only measurement with a Youden's J index above the discriminative threshold value of 0.5 that confirms a useful clinical result [33,62] (Table 2).

Table 2. Sensitivity, specificity and Youden's J index of traditional vital signs and hemodynamic responses for prediction of the onset of decompensated shock secondary to progressive central hypovolemia. Modified from Convertino et al. [14,22,23,25].

Vital Sign	Sensitivity	Specificity	Youden's 'J' Index
Systolic Blood Pressure	0.80	0.17	0.03
Diastolic Blood Pressure	0.40	0.53	0.07
Mean Blood Pressure	0.60	0.33	0.07
Heart Rate	0.80	0.02	0.18
Stroke Volume	0.60	0.33	0.07
Cardiac Output	0.80	0.02	0.18
Pulse Pressure Variability	0.78	0.69	0.47
Peripheral Capillary Oxygen Saturation (SpO$_2$)	0.60	0.00	0.40
Deep Muscle Oxygen Saturation (SmO$_2$)	0.65	0.63	0.28
Compensatory Reserve	0.84–0.87	0.78–0.86	0.62–0.73

Note: For Youden's Index, a value of 1 represents a perfect diagnostic test, while a value of 0 represents a test with poor diagnostic accuracy. Stroke volume (SV), systolic, diastolic and mean blood pressures were measured by finger photoplethysmograpy; heart rate (HR) was measured by standard electrocardiogram; cardiac output was calculated as SV times HR; Pulse pressure variability and SpO$_2$ was measured with standard pulse oximetry; SmO$_2$ was measured with near-infrared spectroscopy; compensatory reserve was measured by pulse oximetry.

The performance of standard vital signs and hemodynamic measurements to provide an early and accurate prediction for onset of decompensated shock can also be assessed with comparisons of sensitivity and specificity calculated using the Area Under the Curve (AUC) Receiver Operating Characteristic (ROC) statistical analysis. Figure 4 provides ROC AUC comparisons of CRM with various hemodynamic (top panel), metabolic (middle panel), and autonomic cardiac (bottom panel as represented by metrics of heart rate variability and complexity) responses. The ROC AUC data in Figure 4 are based on human data generated from experimentally-controlled progressive reductions in central blood volume using the LBNP hemorrhage model [38,39,54,56,57,63]. Similar results have been reported from experiments involving controlled hemorrhage in humans [25,61,64]. These latter data corroborate the results presented in Table 2 that arterial waveform feature analysis provides a monitoring technology with the greatest ability for early and accurate prediction for the onset of decompensated shock.

Optimal management of significant traumatic hemorrhage and other compromising clinical conditions is often delayed by failure to recognize a medical crisis due to the current reliance on traditional vital signs and/or other standard physiological measures that represent a limited assessment of a totally integrated compensatory response [22,24–29,54,61,64]. In this regard, the value of monitoring the arterial waveform morphology for early detection of a clinical crisis using a CRM algorithm has been well documented during actual controlled human hemorrhage in the laboratory setting [14,22,25,38–42,44,50,52,53,61,64,65], and translated to early recognition of hypovolemia and hypotension when used by first responders during simulated emergencies training exercises [66,67], and in hospital critical care settings [20,21,43,45–47,49,51,60,68–72]. The comparative data regarding sensitivity, specificity and diagnostic accuracy of various monitoring technologies presented in this review provide compelling support for the notion that the development of wearable sensors must include an ability to capture analog signals that allow for continuous real-time analysis of changes in features of the analog arterial waveform. It should be recognized that a functional FDA-cleared monitoring system with the CRM algorithm integrated into a standard finger pulse oximeter has been developed and tested [20,22,30,69]. However, such technology has proven to provide limited information to the clinical caregiver about patient status because of unstable positioning and movement artifact. In this regard, we use the following sections of this review to emphasize the need for developing new wearable sensor technologies that can be integrated with the established CRM algorithm in order to advance vital sign monitoring for emergency critical care.

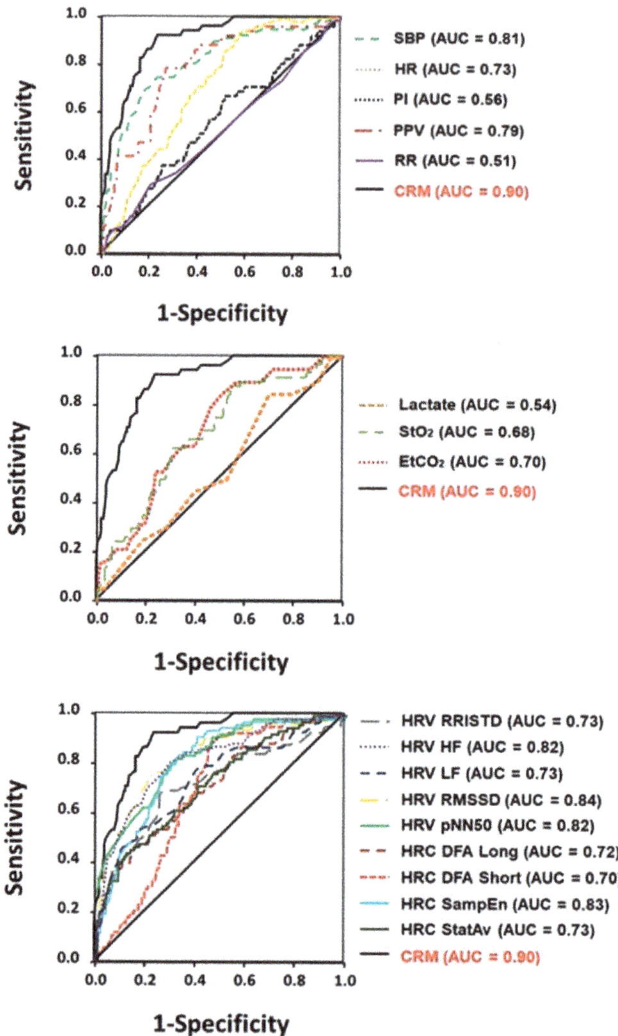

Figure 4. ROC AUC comparisons for prediction of onset of decompensated shock between measures of compensatory reserve (CRM) and standard vital signs (**top** panel), metabolic metrics (**middle** panel), and autonomic nervous system responses measured by indices of heart rate variability (HRV) and complexity (HRC) (**bottom** panel). SBP, systolic blood pressure; HR, heart rate; PI, perfusion index; PPV, pulse pressure variability; RR, respiratory rate; StO_2, tissue oxygen saturation; $EtCO_2$, end-tidal carbon dioxide; RRISD, R-to-R interval standard deviation; HF, high frequency; LF, low frequency; RMSSD, root mean square standard deviation; pNN50, percentage of RRI that vary by at least 50 ms; DFA detrended fluctuation analysis; SampEn, sample entropy; StatAv, stationarity.

4. Current Sensor Technology

4.1. Arterial Waveform Measurement Modalities Amenable to Wearable Technology: Obtaining Reliable High Signal-to-Noise Features

The original studies establishing the basis for CRM [30] used arterial waveforms measured by volume-clamping based continuous blood pressure measurement technology (i.e., Finapres) [73]. Such arterial waveforms have been demonstrated to accurately represent corresponding peripheral blood pressure waveforms obtained using arterial lines [74], and are thereby considered to be a reference standard for non-invasive continuous blood pressure measurement. While the volume-clamping technique is quite accurate at acquiring analog arterial pressure waveforms, the system required is expensive, large, and power-hungry, and thus unsuitable for point-of-care settings. Accordingly, to facilitate translation of CRM outside the lab, investigators have explored other techniques for obtaining analog arterial waveforms that *resemble* blood pressure waveforms; namely, the most commonly employed signal has been the photoplethysmogram (PPG) [39].

Example sensing modalities that provide arterial pressure waveforms (or analogs) that are directly amenable to CRM are summarized in Figure 5. Figure 5a shows volume-clamping based finger cuff BP measurement (i.e., Finapres), which uses an LED and photodiode (PD) to capture the blood volume as a function of time in the finger, and uses a servo controller to modify cuff pressure (P_{cuff}) dynamically to set the finger blood volume to a constant level. The output pressure required from the controller is thus the hemodynamic pressure inside the artery, and a waveform can be outputted representing the continuous arterial BP signal. The measurement can only be obtained from the finger.

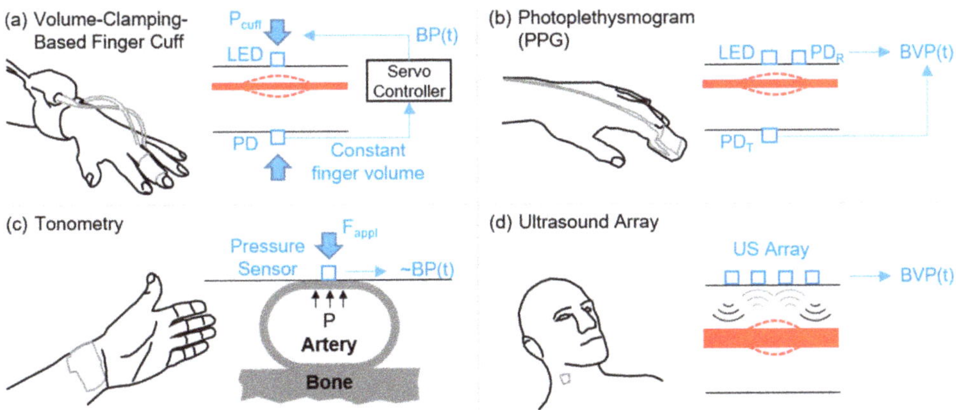

Figure 5. Illustration of different waveforms capture techniques that can provide a blood pressure (BP(t)) or blood volume pulse (BVP(t)) signal. (**a**) Volume-clamping based finger-cuff BP measurement (i.e., Finapres). (**b**) Photoplethysmography (PPG) based blood volume pulse measurement. (**c**) Tonometry based arterial pulse measurement. Image created based on Lee and Nam [75]. (**d**) Ultrasound array based arterial pulse measurement. Image created based on Wang and Xu [76].

Note that the approaches besides volume-clamping based BP measurement would result in waveform characteristics that would differ from the existing library of LBNP based arterial waveform measures used for CRM. Thus, a small data collection of approximately ten subjects may be needed with the new modality such that transfer learning or fine tuning methods for retraining the algorithm can be implemented. Following such methodology, the existing database can still be leveraged with the new sensing modality to yield accurate CRM results.

4.1.1. PPG Signals

Figure 5b shows PPG measurement, which hinges on the acquisition of the BVP waveform, by illuminating a tissue volume with an LED and measuring the transmitted light through the tissue with a PD (PDT) or the light reflected back from the tissue volume with a PD (PDR) [53,77]. The measurement is most commonly obtained from the finger in transmissive mode, but in reflective mode can be measured from other well perfused sites on the body (e.g., forehead, forearm, wrist). PPG is the basis for pulse oximeters, used ubiquitously for measuring arterial oxygen saturation in clinical settings. With each heartbeat, the volume of arterial blood in the tissue being illuminated decreases during diastole and increases during systole, and thus the light passing through the tissue is brighter (diastole) and dimmer (systole) at the photodiode. Since the volumetric expansion and contraction of the arteries is dependent on pulse pressure and arterial compliance, the PPG waveform closely resembles the underlying arterial blood pressure waveform in shape. While PPG waveforms are captured on commercially available pulse oximeter instrumentation, such waveforms may not be reliable for CRM since the PPG signals are heavily filtered and processed [77]. PPG can be measured in both transmissive and reflectance mode: for transmissive mode operation, the LED and photodiode are on opposite sides of the tissue (typically the earlobe, fingertip, or toe); for reflectance mode operation, the LED and photodiode are adjacent to one another on the same side of the tissue, and thus the locations for measurement can theoretically be anywhere on the body with sufficient perfusion (e.g., forehead, forearm, chest, and wrist). The main disadvantages of reflectance-mode PPG are that the signal quality is lower [78], the measurement varies with positioning and the distance between the LED and the photodiode, and the signal is more affected by motion artifacts [79,80]. Recent developments in device fabrication have allowed PPG sensing systems to be flexible and skin-interfaced for comfortable use in long-term care scenarios [81,82]. Soft and stretchable optoelectronics sensing for transmissive PPG measurement was demonstrated by Biswas, et al. [83]. An interesting approach not requiring an LED but rather using ambient light for PPG sensing was demonstrated by Han, et al.; with this approach, PPG signals with distinguishable heartbeat peaks were recorded and corresponding pulse oximetry readings were obtained [84].

4.1.2. Tonometry Signals

Figure 5c shows tonometry measurement, which involves the application of a force to flatten (or applanate) an artery with a given applanation force (F_{appl}), and a pressure sensor applied to the skin above the artery then records the time varying fluctuations in pressure applied by the blood on the arterial wall [85,86]. With perfect applanation, this pressure waveform (BP(t)) is exactly equal to arterial pressure; however, in practice, applanation is usually imperfect and thus the waveform simply resembles BP. The most common measurement site is the radial artery. The advantage in tonometer measurements as compared to PPG is that substantially lower power consumption is required [87]—PPG employs active sensing where light is delivered to the tissue and then the resultant transmitted or reflected light level is detected; tonometry is a passive measurement where a transducer simply records the distension of the arterial wall. However, the major disadvantage in tonometry as compared to PPG is that the measurement is highly dependent on the location, and the transducer must be reliably placed over a superficial artery.

4.1.3. Wearable Ultrasound

Figure 5d shows ultrasound array based measurement, which uses an array of ultrasound transducers in a flexible form factor placed on the skin to measure arterial diameter changes versus time for a large artery (e.g., the carotid artery). Changes in arterial diameter correspond to the BVP, but are measured from a deeper artery as compared to PPG or tonometry, and thus may be less affected by vasoconstriction. A common measurement site is the carotid artery. Recent work has demonstrated that a blood pressure waveform can be measured from the surface of the skin based on this principle using a nano-engineered flexible ultrasound array [88]. The device acquires time-varying changes in blood vessel diameter, which are then mapped to an estimate of the underlying blood pressure waveforms. By employing ultrasound to measure this pulsating blood vessel diameter, the device can focus on larger arteries, namely the carotid, which are deeper under the skin than PPG- or tonometer-based approaches can access. Accordingly, reduced sensitivity to sensor positioning has been demonstrated as compared to tonometry, and accurate extraction of arterial pressure waveforms has been achieved [88]. Note that this approach requires calibration to acquire the absolute blood pressure values (i.e., systolic, diastolic, and mean arterial pressure), but the waveforms measured are likely the closest to the underlying blood pressure waveforms of the three prior modalities discussed here. An additional concern that should be noted with this approach is that the detection of the artery may require manual positioning and/or image annotation in broad use; however, the array of transducers employed on the device may limit such a need for expert annotation.

4.1.4. Cardio-Mechanical Vibrations

While CRM to date has focused on arterial pulse waveforms measured peripherally, there have also been studies employing cardiogenic vibration signals as an index of hypovolemia based on machine learning techniques in both human subjects (LBNP) [89] and animal models [90]. Note that since these measurements to do not directly yield an arterial pulse waveform, they are not depicted in Figure 5 to avoid confusion. Cardiogenic vibration signals include the seismocardiogram (SCG) and ballistocardiogram (BCG), both of which originate from the vibrations of the chest (SCG) or whole body (BCG) in response to the ejection of blood from the heart and movement of blood through the vasculature [91]. SCG and BCG signals can be measured accurately with inexpensive and commercially available sensors [92,93], and have been demonstrated to be reliable even in the presence of movement [94,95]. As with the other sensing modalities described above, soft, conformal patch based sensing of SCG signals is also possible: Liu, et al. describe an epidermal sensing system for providing mechano-acoustic measurements of cardiovascular health, including heart sounds and SCG signals [96]. Machine learning based analyses performed on these waveforms demonstrated that high quality estimation of blood volume status (analogous to CRM) could be obtained in a pig model of hemorrhage [90]. Importantly, for persons suffering polytraumatic injuries who may not have an available digit or ear, and may have extensive vascular damage that could reduce PPG waveform quality due to increased wave reflections, such cardiogenic vibrations may provide an alternative waveform for CRM-based volume status assessment.

4.1.5. Other Emerging Wearable Sensing Devices

The field of wearable sensing has seen a myriad of new devices over the past several years, driven by the use of new materials and fabrication approaches, developments in chemical sensing, and the advent of soft, flexible, and stretchable electronics. These new devices promise to deliver comfortable and high-performance sensing of cardiovascular health parameters with thin, flexible, and stretchable mechanical footprints that resemble the properties of human skin. Emerging technologies of interest also include biodegradable sensors such as the one described in Boutry, et al., for tonometry-based pulse signature sensing [97], and combined chemical/electrophysiological hybrid biosensing systems such as the one presented in Imani, et al. [98]. Additionally, while not discussed in detail here, wearable sensing systems measuring impedance plethysmogram waveforms [99], or magnetic inductance based cardiac waveforms [100], may also be employed.

Table 3 provides a comparison of state-of-the-art sensing technologies for arterial pulse waveform analogs, including summarizing the principle of operation, the typical locations on the body where the signals are captured, and the advantages and disadvantages of each method for application to CRM.

Table 3. Comparison of Sensing Technologies for Arterial Pulse Waveform Analogs.

Sensing Modality	Principle of Operation	Typical Location(s)	Advantages	Disadvantages
PPG	Optical sensing of blood volume changes in a small volume of tissue	Transmissive: Finger, Earlobe, Toe Reflective: Wrist, Forehead, Forearm	Waveform resembles arterial pressure curves; signal quality is typically high; well-established sensing modality and already used in many clinical settings (i.e., pulse oximetry)	Signal originates mainly from the cutaneous vasculature and thus is affected by hypoperfusion (peripheral vasoconstriction); reflective PPG is more convenient in terms of placement but suffers from motion artifacts and placement-based variability in signal shape; requires substantial current consumption (active sensing)
Tonometry	Force/pressure sensing of arterial wall displacement at the surface of the skin	Wrist (Radial Artery)	With applanation of the artery, captures true blood pressure waveform and does not require calibration; low power measurement since it is a passive sensing approach	Applanation is challenging in practice and not reliable; measurement depends highly on placement; coupling to a superficial artery is needed
Ultrasound-Based BP	Ultrasound sensing of arterial diameter changes in deeper/larger arteries	Neck (Carotid Artery)	Measurements can be obtained from deeper arteries (e.g., carotid) and thus are less affected by hypoperfusion and/or vasoconstriction; arrayed sensing approach may reduce the variability in signal shape due to sensor positioning	Active measurement which requires substantial power consumption to deliver ultrasound energy to the body and process the resultant signals; may require manual approaches to annotating images
Cardiogenic Vibration	Mechanical vibration sensing of blood movement through vasculature	Chest (Sternum)	Passive measurements that can be captured non-intrusively with sensors on the chest; represent more central cardiac activity since the origination is from cardiac vibrations rather than peripheral blood volume pulse; minimal affects due to peripheral vasoconstriction	Not a direct arterial pressure waveform analog; requires coupling to the chest with an adhesive; may be sensitive to positioning of the sensor on the body

4.2. Mitigating the Effects of External Vibrations and Motion Artifacts

In point-of-care settings, sensing systems for obtaining arterial pulse waveforms often encounter external vibration or motion related artifacts (e.g., battlefield settings or civilian patient transport). These artifacts can greatly impact the quality of the waveforms that are measured, and result in errors in the computation of clinically relevant information (e.g., CRM). External vibrations from transport vehicles during en route care, for example, can be quite large (e.g., on an ambulance or helicopter). Motion artifacts will always be present in the measured signals, unless the patient is unconscious, and may result from either whole-body movements or even more subtle sources such as respiration, talking, or coughing. There are many approaches for mitigating the effects of vibration and motion artifacts in arterial waveform measurements, but the most common techniques involve: (1) improving the signal quality at the source as much as possible, (2) providing auxiliary sensors to detect and cancel motion artifacts from the measured waveforms, and (3) quantifying signal quality on a beat-by-beat basis to facilitate rejection of lower quality waveforms from the subsequent data analysis. Figure 6 shows an example of BP and PPG signals (green, red, and infrared wavelength) captured from a representative subject using a wearable watch technology [101] (a) while at rest (b) and following vigorous exercise with example motion artifacts (c).

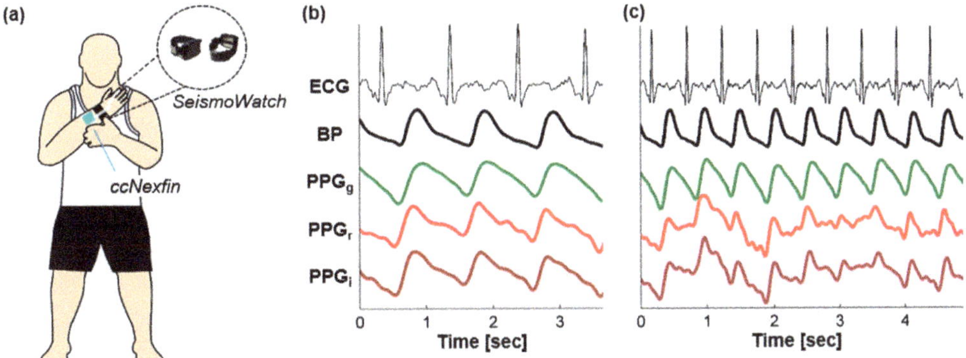

Figure 6. Example PPG signals taken together with electrocardiogram (ECG) and BP measurements as a reference for comparison. (**a**) The signals were obtained by the Georgia Tech SeismoWatch hardware as described in Ganti, et al. [102]. The BP waveforms shown for comparison were obtained with the ccNexfin volume-clamping based finger cuff BP system (Edwards Lifesciences). (**b**) Signals measured from a subject at rest. Note that the PPG waveforms closely resemble the BP waveforms in shape, with the red and IR (PPG$_r$ and PPG$_i$, respectively) containing many of the same characteristics expected in an arterial pulse waveform, while green (PPG$_g$) appears to be a smoothed version of the BP waveform. (**c**) Signals measured from the same subject following heavy exercise with motion artifacts corrupting the waveforms. The red and IR signals are corrupted heavily while the green PPG signal quality remains high.

4.2.1. Improving the Signal Quality at the Source

For reflectance-mode PPG signals, signal quality is optimized at the source against motion artifacts through the use of green wavelengths rather than red or infrared (IR) [79,102,103]; green penetrates less deeply into the skin, and thus is less attenuated through the forward and backward path through the tissue. Providing non-zero contact pressure between the PPG sensor and the skin can also increase the amplitude of the measured waveforms [53,104]. Specifically, the PPG amplitude is maximized when the contact pressure is equal to the mean arterial pressure (i.e., the transmural pressure is zero). Thus, to reduce the impact of motion artifacts, green wavelengths can be employed for PPG detection, and a non-zero contact pressure can be applied between the sensor and the skin to optimize signal level. The waveforms shown in Figure 6c visually demonstrate this relationship between wavelength and resultant PPG signal quality during motion artifacts. While the red and IR PPG signals are quite heavily affected by the motion artifacts, the green PPG signal quality remains high. Nevertheless, note that many of the key waveform features captured by the red and IR PPG are missing in the green PPG signal due to the fact that the green signal captures primarily the superficial cutaneous vasculature while red and IR penetrate deeper into the skin.

For tonometry-based arterial pulse waveform measurements, optimizing signal quality at the source fundamentally requires robust coupling between the superficial artery and the sensor. Tonometry requires a backing force such that the sensor remains consistently in contact with the arterial wall throughout the measurement duration. Thus, a strap is typically used to provide such backing force, for example for radial artery tonometry, and the tightness of the strap must be optimized to be high enough such that the sensor remains in contact with the artery but not high enough to occlude the artery [105]. To reduce the variability due to sensor placement, arrayed sensors are also often used for tonometry based recordings [106]. The sensing system can thus be placed over the palmar aspect of the wrist near the radius bone, and software based approaches can be used to find the waveform with the highest signal quality from the array of sensors.

Techniques for optimizing wearable ultrasound array based arterial pulse waveforms are not yet well understood since the measurement modality is relatively new. However, from an intuitive perspective it is likely that the ability to accurately place the ultrasound array in the proximity of the artery from which measurements will be taken (e.g., the carotid artery), and the coupling between the sensor array and the skin (likely requiring acoustic matching such as ultrasound gel), will play important roles in ensuring high quality waveforms are obtained.

For cardiogenic vibration signals, there are also several aspects that must be considered to optimize signal quality at the source. First, the sensing system should use a sensor with sufficiently low noise floor to capture the micro-vibrations. In the case of SCG signals for example, only accelerometers with input-referred noise of 50 $\mu g_{rms}/\sqrt{Hz}$ or lower should be used. The standard accelerometers deployed on wearable sensing systems and smartphones for inertial measurement have much higher noise than this, with values typically in the 150–300 $\mu g_{rms}/\sqrt{Hz}$ range. Second, leveraging the information from all three axes of the SCG signal, or even including rotational components (i.e., gyrocardiography) as captured with a gyroscope, has been demonstrated to yield greater information than the dorso-ventral axis alone [107]. Third, and perhaps most importantly, the sensor should be rigidly adhered to the body such that movement of the person wearing the sensor does not lead to detachment or other major mechanical disturbances.

4.2.2. Providing Auxiliary Sensors to Detect and Cancel Motion Artifacts

A commonly-used technique for reducing the impact of motion artifacts on PPG signals is the inclusion of an auxiliary accelerometer to detect and provide digital subtraction of motion artifacts [108–110]. The captured acceleration signal provides a noise reference that can be used via adaptive noise cancellation or other signal processing approaches to remove such artifacts from the PPG signal. An alternative approach to reducing motion artifact influence on wearable cardio-mechanical signals leverages auxiliary sensing to capture other signals of cardiovascular origin, namely the electrocardiogram (ECG) [111,112]. Subsequently, rather than removing motion artifacts, the signal strength itself can be bolstered. While the authors are not aware of such auxiliary sensor-based methods for increasing robustness to motion artifacts in tonometry and ultrasound-based arterial waveform capture modalities, intuitively such methods should be directly applicable to these modalities as well. The fundamental approach of either providing a noise reference for noise cancellation or a timing reference for ensemble averaging or otherwise strengthening the signal characteristics are valid for these modalities similarly as for PPG signals.

For cardiogenic vibration waveforms, several approaches have been demonstrated in the existing literature for detecting and cancelling artifacts due to motion or external vibration. Auxiliary sensors for detecting or cancelling motion artifacts from BCG signals include foot electromyogram (EMG) sensing to determine periods of elevated motion as well as external geophone based recordings of floor vibrations for subsequent cancellation [113,114]. Furthermore, signal enhancement using concurrent ECG signals for ensemble averaging, synchronized moving averaging, and otherwise beat segmentation is standard practice.

4.2.3. Quantifying Signal Quality for Rejecting Lower Quality Waveforms

A third approach that can be leveraged to mitigate the effects of external vibrations and motion artifacts on arterial pulse waveforms is the automatic quantification of signal quality on a beat-by-beat basis. Such signal quality assessment is an important tool towards quantifying when the waveforms should be inputted to subsequent machine learning steps (e.g., CRM computation) or, alternatively, when waveform segments should be rejected. Signal quality indices have thus been developed for PPG and cardio-mechanical signals, and have been validated in recent literature [115–117]. The challenge in such algorithms is that both the signal (of cardiac origin) and the noise are non-stationary, and there is substantial variability in signal shape across subjects and also sensor locations. Thus, conventional approaches such as matching the morphology of measured PPG (or tonometry, ultrasound, cardiogenic vibration signals, etc.) to previous recordings or a database of recordings is not an appropriate technique. Rather, waveform matching must be accomplished using techniques such as dynamic time warping (DTW) [118], which allow for stretching of each beat against the templates with which the beat is compared. DTW-based approaches have demonstrated promise for arterial pulse signals [116]. The establishment of such automated techniques for signal quality assessment—as compared to manual annotation which has been employed in many studies in the existing literature—will represent an important step towards facilitating translation of these sensing approaches to point-of-care settings. Note that, whenever possible, techniques for improving signal quality should be employed rather than techniques for only assessing signal quality. However, in practical settings, many sources of artifacts, noise, and interference corrupting physiological measurements cannot be completely attenuated by signal capture optimization, nor can they be completely removed by auxiliary sensors and associated noise reduction algorithms; thus, the ability to detect and remove low quality events is a key element in delivering robust and reliable CRM outputs to caregivers for subsequent clinical decision making.

4.3. Eliminating the Need for Baseline Measures/Calibration

The use of wearable sensors for CRM-based hypovolemia assessment in field settings may not allow for baseline data to be obtained; for example, if one envisions a person injured in a major car accident, an emergency medical technician (EMT) may simply apply a wearable patch or system to the person when arriving on the scene after exsanguination has begun. Accordingly, algorithms for quantifying compensatory reserve based on machine learning should be globalized rather than designed in a patient-specific manner (see Figure 3). Features leveraged by the algorithm should thus be based on relative measures (e.g., timing intervals, variability measures, etc.) rather than absolute measures (e.g., absolute amplitude of the signal). Moreover, machine learning algorithms should be trained using leave-one-subject-out cross-validation (LOSO-CV) rather than n-fold CV, with at least one subject deliberately left out of the training set such that the algorithm focuses on global trends in the sensed waveforms. Finally, since sensor placement can impact the shape of waveforms measured for many reflective PPG [79], tonometry, ultrasound-based blood pressure, and cardiogenic vibration signals [117], such placement-dependent changes should be thoroughly quantified, and methods for harnessing underlying dynamics should be leveraged as compared to features that require manual annotation [118].

4.4. Real-Time Measurements and Processing for Display

An important consideration is how to display the resultant information derived from the arterial waveforms to the physician or caregiver. One option is to provide a dashboard type display with perhaps a single numerical value indicating the compensatory status of the person (i.e., a CRM value). Another option might be a red, yellow, or green indicator to provide information regarding the clinical decisions to be made during triage (Figure 3). An exciting opportunity exists in the pairing of the volume status information delivered through the automatic analysis of the arterial pulse waveforms with autonomous critical care systems for combat casualty care. Scientists in the academic and commercial domain are conducting research designed to develop systems and methods for providing fluids autonomously to combat casualties based on physiological data [119–122]; providing more in-depth measurements of volume status beyond traditional vital signs to such systems may yield improved results in managing fluid for hemorrhaging patients or casualties. As different applications and use cases emerge, it will be important to determine what processing will be applied at what stage in the system. For example, in one implementation the signals may be wirelessly transmitted from the wearable sensing system to a local smartphone, tablet, laptop, or other dedicated receiver, at which point algorithms may be implemented on that receiver device to output a CRM to be displayed to the caregiver. Another implementation that is possible is to incorporate the CRM machine learning algorithm into the wearable hardware itself (i.e., computing on the edge), in which case the CRM value itself may be transmitted wirelessly or a readout may be provided on the hardware itself. Regardless of where in the signal chain the processing is implemented, it will be necessary to consider security and privacy concerns of the patient, as well as power consumption and associated battery life on the wearable hardware itself.

4.5. Electronic Documentation in the Prehospital Setting

The ability to collect and analyze large quantities of data from trauma patients, particularly in austere prehospital settings such as the battlefield, hinder the potential for understanding and improving clinical process and performance [123]. In situations where battery life must be extended for as long as possible, or when wireless transmission is otherwise not feasible, data storage locally on the sensing system may be desired and implemented using micro secure digital (microSD) cards or other non-volatile memory on board the system. The advantage to such local storage of all physiological waveforms is that a detailed record can be kept of the data for subsequent analysis and/or evaluation of the treatment approaches employed. Data extracted from all patients could then be used to retrospectively determine which approaches were most successful, and care can then be optimized accordingly with this evidence. In some instances, the amount of data being stored may be quite large, and may necessitate compressed sensing approaches prior to digitization [124,125]. However, in most cases—since physiological signals such as the PPG are typically of low bandwidth (<100 Hz)—direct digitization and storage of data are feasible for many weeks of continuous recording.

4.6. Military Perspectives and Implications

In December 2013, the Director of the former Directorate of Combat and Doctrine Development (currently the Capability Development Integration Directorate) signed a 'Requirements Adjudication Team' memorandum that documented a military medical requirement for the measurement of compensatory reserve. The Committee on Tactical Combat Casualty Care reaffirmed this requirement by recommending "continued development and expedited fielding of technologies (such as the compensatory reserve) that enable prehospital combat medical personnel to better evaluate the need for and the adequacy of fluid resuscitation" [55]. As military missions of the future will be performed in complex multi-domain operations (MDO) and/or involve large scale combat operations (LSCO) with a possibility of limited air superiority, delays in early and rapid medical evacuation in addition to mass casualty scenarios will require individualized triage decision support that will prove critical for successful execution of prolonged field care. In the military setting, warfighters could wear a sensor embedded on a wrist watch (e.g., Figure 6) or as part of their fighting ensemble system so that the clinical status of injured casualties could support continuous hands-free documentation by a combat medic using a remote monitoring device (e.g., phone). Since previous research has also identified the CRM with the capability to track physical and physiological performance [70,126,127], a military wearable sensor system that integrates the continuous monitoring of CRM could be used by unit commanders as a real-time metric of performance readiness (e.g., manage impact of heat strain and/or dehydration) as well as its use for optimizing combat casualty care of warfighters in austere battlefield settings.

4.7. Future Directions

Future work is required to collectively advance the vision of enabling CRM-based assessment of hypovolemia in field settings. Wearable sensing systems are needed with minimally obtrusive form factors facilitating the accurate measurement of arterial pulse or cardio-mechanical waveforms outside of laboratory settings. Such systems should likely employ multi-modal sensing approaches: for example, PPG sensing can be combined with tonometry and/or cardiogenic vibration sensing to ensure that if one modality experiences artifacts from motion or other confounding variables, the other modality might still accurately capture cardiac signatures. The physiological origins of the signals being measured, and the manner in which confounding variables such as environmental factors (e.g., ambient temperature), arrhythmias, other cardiovascular disease conditions, and high body mass index of the patient may impact the algorithms and/or sensor design should be investigated further. The specific features and signal modalities that might offer the most salient information regarding volume status should continue to be studied through LBNP, heat stress/dehydration, and other hypovolemia inducing protocols. These wearable sensing systems must be paired with state-of-the-art machine learning algorithms to reduce noise and interference, automatically assess signal quality, and output a reliable and robust indication of CRM. Designers of such hardware, firmware, and software required for this framework should collaborate closely with subject matter experts such as medical professionals and EMTs such that the user interface and display offered to these professionals provides the information needed for rapid decision making in the challenging environment of prehospital trauma care. Finally, extensive validation of these technologies as a whole must be conducted to ensure that the performance of all components of the overall system are sufficiently robust to obtain regulatory approval and ultimately improve outcomes.

5. Conclusions

As technology advances to facilitate the emergence of autonomous medical treatment systems as well as early and accurate diagnosis and triage, the incorporation of sensors capable of supporting measurements of CRM can ensure that patients who require emergency medical care (e.g., civilian trauma patients or wounded service members) receive appropriate treatment interventions, even when medical personnel are not available. As such, the development and availability of a single advanced monitoring system that includes wearable sensors capable of capturing analog arterial waveforms and integrating them with application of machine-learning algorithms (i.e., artificial intelligence) can provide clinical and/or performance decision-support with the goal of optimizing health, safety and wellbeing in prehospital and emergency room settings. In addition to offering robust performance, human factors aspects of the sensing system design must be prioritized such that both the hardware and clinician-facing displays seamlessly integrate into the workflow, making it easier for decisions to be made in time-critical, challenging situations. Finally, such systems and associated algorithms as described in this review paper may be applied to the diagnosis or management of other cardiovascular conditions, such as heart failure management.

Author Contributions: Conceptualization, V.A.C., S.G.S., E.K.W., S.C., M.N.S., and O.T.I.; writing—original draft preparation, V.A.C., S.G.S., O.T.I.; writing—review and editing, V.A.C., S.G.S., E.K.W., S.C., M.E.S., M.J.T., M.N.S., and O.T.I. All authors have read and agreed to the published version of the manuscript.

Funding: This work was funded by the U.S. Army Medical Research and Development Command and the Congressionally Directed Medical Research Program (award number DM180240).

Acknowledgments: The authors thank Aiyana Helme for her assistance in preparing the manuscript and Venu Ganti for his assistance with figure preparation.

References

1. Charlari, E.; Intas, G.; Stergiannis, P.; Vezyridis, P.; Fildissis, G. The importance of vital signs in the triage of injured patients. *Crit. Care Nurs. Q.* **2013**, *35*, 292–298. [CrossRef] [PubMed]
2. Booth, J. A short history of blood pressure measurement. *Proc. R. Soc. Med.* **1977**, *70*, 793–799. [CrossRef] [PubMed]

3. Schauer, S.G.; Naylor, J.F.; April, M.D.; Fisher, A.D.; Cunningham, C.W.; Fernandez, J.R.D.; Shreve, B.P.; Bebarta, V.S. Prehospital resuscitation performed on hypotensive trauma patients in afghanistan: The prehospital trauma registry experience. *Mil. Med.* **2019**, *184*, e154–e157. [CrossRef] [PubMed]

4. Naylor, J.F.; Fisher, A.D.; April, M.D.; Schauer, S.G. An analysis of radial pulse strength to recorded blood pressure in the Department of Defense Trauma Registry. *Mil. Med.* **2020**. [CrossRef]

5. Holcomb, J.B.; Salinas, J.; McManus, J.M.; Miller, C.C.; Cooke, W.H.; Convertino, V.A. Manual vital signs reliably predict need for life-saving intervention in trauma patients. *J. Trauma* **2005**, *59*, 821–829. [CrossRef]

6. McManus, J.; Yershov, A.L.; Ludwig, D.; Holcomb, J.B.; Salinas, J.; Dubick, M.A.; Convertino, V.A.; Hinds, D.; David, W.; Flanagan, T.; et al. Radial pulse character relationships to systolic blood pressure and trauma outcomes. *Prehospital Emerg. Care* **2005**, *9*, 423–428. [CrossRef]

7. Keenan, S.; Riesberg, J.C. Prolonged field care: Beyond the "Golden Hour". *Wilderness Environ. Med.* **2017**, *28*, S135–S139. [CrossRef]

8. Ball, J.A.; Keenan, S. Prolonged field care working group position paper: Prolonged field care capabilities. *J. Spec. Oper. Med.* **2015**, *15*, 76–77.

9. Bickell, W.H.; Wall, M.J., Jr.; Pepe, P.E.; Martin, R.R.; Ginger, V.F.; Allen, M.K.; Mattox, K.L. Immediate versus delayed fluid resuscitation for hypotensive patients with penetrating torso injuries. *N. Engl. J. Med.* **1994**, *331*, 1105–1109. [CrossRef]

10. Griggs, J.E.; Jeyanathan, J.; Joy, M.; Russell, M.Q.; Durge, N.; Bootland, D.; Dunn, S.; Sausmarez, E.D.; Wareham, G.; Weaver, A.; et al. Mortality of civilian patients with suspected traumatic haemorrhage receiving pre-hospital transfusion of packed red blood cells compared to pre-hospital crystalloid. *Scand. J. Trauma Resusc. Emerg. Med.* **2018**, *26*, 100. [CrossRef]

11. Huang, G.S.; Dunham, C.M. Mortality outcomes in trauma patients undergoing prehospital red blood cell transfusion: A systematic literature review. *Int. J. Burns Trauma* **2017**, *7*, 17–26. [PubMed]

12. Bellamy, R.F. The causes of death in conventional land warfare: Implications for combat casualty care research. *Mil. Med.* **1984**, *149*, 55–62. [PubMed]

13. Eastridge, B.J.; Hardin, M.; Cantrell, J.; Oetjen-Gerdes, L.; Zubko, T.; Mallak, C.; Wade, C.E.; Simmons, J.; Mace, J.; Mabry, R.; et al. Died of wounds on the battlefield: Causation and implications for improving combat casualty care. *J. Trauma* **2011**, *71*, S4–S8. [CrossRef] [PubMed]

14. Convertino, V.A.; Koons, N.J.; Suresh, M. The physiology of human hemorrhage and Compensation. *Comp. Physiol.* **2020**, in press.

15. April, M.D.; Becker, T.E.; Fisher, A.D.; Naylor, J.F.; Schauer, S.G. Vital sign thresholds predictive of death in the combat setting. *Am. J. Emerg. Med.* **2020**. [CrossRef] [PubMed]

16. Eastridge, B.J.; Salinas, J.; McManus, J.G.; Blackburn, L.; Bugler, E.M.; Cooke, W.H.; Convertino, V.A.; Wade, C.E.; Holcomb, J.B. Hypotension begins at 110 mm Hg: Redefining "hypotension" with data. *J. Trauma* **2007**, *63*, 291–297. [CrossRef] [PubMed]

17. Koons, N.J.; Nguyen, B.; Suresh, M.R.; Hinojosa-Laborde, C.; Convertino, V.A. Tracking DO2 with compensatory reserve during whole blood resuscitation in baboons. *Shock* **2020**, *53*, 327–334. [CrossRef]

18. Convertino, V.A.; Koons, N.J. The compensatory reserve: Potential for accurate individualized goal-directed whole blood resuscitation. *Transfusion* **2020**, *60*, S150–S157. [CrossRef]

19. Thiele, R.H.; Nemergut, E.C.; Lynch, C. The physiologic implications of isolated alpha(1) adrenergic stimulation. *Anesth. Analg.* **2011**, *113*, 284–296. [CrossRef] [PubMed]

20. Convertino, V.A.; Wampler, M.R.; Johnson, M.; Alarhayem, A.; Le, T.D.; Nicholson, S.; Myers, J.G.; Chung, K.K.; Struck, K.R.; Cuenca, C.; et al. Validating clinical threshold values for a dashboard view of the compensatory reserve measurement for hemorrhage detection. *J. Trauma Acute Care Surg.* **2020**, *89*, S169–S174. [CrossRef]

21. Convertino, V.A.; Grudic, G.; Mulligan, J.; Moulton, S. Estimation of individual-specific progression to impending cardiovascular instability using arterial waveforms. *J. Appl. Physiol.* **2013**, *115*, 1196–1202. [CrossRef] [PubMed]

22. Convertino, V.A.; Wirt, M.D.; Glenn, J.F.; Lein, B.C. The compensatory reserve for early and accurate prediction of hemodynamic compromise: A review of the underlying physiology. *Shock* **2016**, *45*, 580–590. [CrossRef] [PubMed]

23. Convertino, V.A.; Schiller, A.M. Measuring the compensatory reserve to identify shock. *J. Trauma Acute Care Surg.* **2017**, *82*, S57–S65. [CrossRef] [PubMed]

24. Convertino, V.A.; Moulton, S.L.; Grudic, G.Z.; Rickards, C.A.; Hinojosa-Laborde, C.; Ryan, K.L. Use of advanced machine-learning techniques for non-invasive monitoring of hemorrhage. *J. Trauma* **2011**, *71*, S25–S32. [CrossRef] [PubMed]

25. Convertino, V.A.; Cardin, S.; Batchelder, P.; Grudic, G.Z.; Mulligan, J.; Moulton, S.L.; MacLeod, D. A novel measurement for accurate assessment of clinical status in patients with significant blood loss: The compensatory reserve. *Shock* **2015**, *44*, 27–32. [CrossRef]

26. Orlinsky, M.; Shoemaker, W.; Reis, E.D.; Kerstein, M.D. Current controversies in shock and resuscitation. *Surg. Clin. N. Am.* **2001**, *81*, 1217–1262. [CrossRef]

27. Wo, C.C.; Shoemaker, W.C.; Appel, P.L.; Bishop, M.H.; Kram, H.B.; Hardin, E. Unreliability of blood pressure and heart rate to evaluate cardiac output in emergency resuscitation and critical illness. *Crit. Care Med.* **1993**, *21*, 218–223. [CrossRef]

28. Bruijns, S.R.; Guly, H.R.; Bouamra, O.; Lecky, F.; Lee, W.A. The value of traditional vital signs, shock index, and age-based markers in predicting trauma mortality. *J. Trauma Acute Care Surg.* **2013**, *74*, 1432–1437. [CrossRef]

29. Parks, J.K.; Elliott, A.C.; Gentilello, L.M.; Shafi, S. Systemic hypotension is a late marker of shock after trauma: A validation study of advanced trauma life support principles in a large national sample. *Am. J. Surg.* **2006**, *192*, 727–731. [CrossRef]

30. Moulton, S.L.; Mulligan, J.; Grudic, G.Z.; Convertino, V.A. Running on empty? The compensatory reserve index. *J. Trauma Acute Care Surg.* **2013**, *75*, 1053–1059. [CrossRef]

31. Hinojosa-Laborde, C.; Howard, J.T.; Mulligan, J.; Grudic, G.Z.; Convertino, V.A. Comparison of comensatory reserve during lower-body negative pressure and hemorrhage in nonhuman primates. *Am. J. Physiol. Regul. Integr. Comp. Physiol.* **2016**, *310*, R1154–R1159. [CrossRef] [PubMed]

32. Hinojosa-Laborde, C.; Shade, R.E.; Muniz, G.W.; Bauer, C.; Goei, K.A.; Pidcoke, H.F.; Chung, K.K.; Cap, A.P.; Convertino, V.A. Validation of lower body negative pressure as an experimental model of hemorrhage. *J. Appl. Physiol.* **2014**, *116*, 406–415. [CrossRef] [PubMed]

33. Youden, W.J. Index for rating diagnostic tests. *Cancer* **1950**, *3*, 32–35. [CrossRef]

34. Convertino, V.A.; Rickards, C.A.; Ryan, K.L. Autonomic mechanisms associated with heart rate and vasoconstrictor reserves. *Clin. Auton. Res.* **2012**, *22*, 123–130. [CrossRef]

35. Engelke, K.A.; Doerr, D.F.; Crandall, C.G.; Convertino, V.A. Application of acute maximal exercise to protect orthostatic tolerance after simulated microgravity. *Am. J. Physiol.* **1996**, *271*, R837–R847. [CrossRef] [PubMed]

36. Convertino, V.A. G-Factor as a tool in basic research: Mechanisms of orthostatic tolerance. *J. Gravit. Physiol.* **1999**, *6*, 73–76.

37. Carter, R., III; Hinojosa-Laborde, C.; Convertino, V.A. Variability in integration of mechanisms associated with high tolerance to progressive reductions in central blood volume: The compensatory reserve. *Physiol. Rep.* **2016**, *4*, e12705. [CrossRef]

38. Howard, J.T.; Janak, J.C.; Hinojosa-Laborde, C.; Convertino, V.A. Specificity of compensatory reserve and tissue oxygenation as early predictors of tolerance to progressive reductions in central blood volume. *Shock* **2016**, *46*, 68–73. [CrossRef]

39. Janak, J.C.; Howard, J.T.; Goei, K.A.; Weber, R.; Muniz, G.W.; Hinojosa-Laborde, C.; Convertino, V.A. Predictors of the onset of hemodynamic decompensation during progressive central hypovolemia: Comparison of the peripheral perfusion index, pulse pressure variability, and compensatory reserve index. *Shock* **2015**, *44*, 548–553. [CrossRef]

40. Chew, M.S.; Aneman, A. Haemodynamic monitoring using arterial waveform analysis. *Curr. Opin. Crit. Care* **2013**, *19*, 234–241. [CrossRef]

41. Convertino, V.A. Blood pressure measurement for accurate assessment of patient status in emergency medical settings. *Aviat. Space Environ. Med.* **2012**, *83*, 614–619. [CrossRef] [PubMed]

42. Convertino, V.A.; Cooke, W.H.; Holcomb, J.B. Arterial pulse pressure and its association with reduced stroke volume during progressive central hypovolemia. *J. Trauma* **2006**, *61*, 629–634. [CrossRef] [PubMed]

43. Davies, S.J.; Vistisen, S.T.; Jian, Z.; Hatib, F.; Scheeren, T.W.L. Ability of an arterial waveform analysis-derived hypotension prediction index to predict future hypotensive events in surgical patients. *Anesth. Analg.* **2020**, *130*, 352–359. [CrossRef] [PubMed]

44. Hametner, B.; Wassertheurer, S. Pulse Waveform Analysis: Is it ready for prime time? *Curr. Hypertens. Rep.* **2017**, *19*, 73. [CrossRef]

45. Hatib, F.; Jian, Z.; Buddi, S.; Lee, C.; Settels, J.; Sibert, K.; Rinehart, J.; Cannesson, M. Machine-learning algorithm to predict hypotension based on high-fidelity arterial pressure waveform analysis. *Anesthesiology* **2018**, *129*, 663–674. [CrossRef]

46. Kendale, S.; Kulkarni, P.; Rosenberg, A.D.; Wang, J. Supervised machine-learning predictive analytics for prediction of postinduction hypotension. *Anesthesiology* **2018**, *129*, 675–688. [CrossRef]

47. Kim, S.K.; Shin, W.J.; Kim, J.W.; Park, J.Y.; Hwang, G.S. Prediction of hyperdynamic circulation by arterial diastolic reflected waveform analysis in patients undergoing liver transplantation. *Blood Press. Monit.* **2016**, *21*, 9–15. [CrossRef]

48. Thiele, R.H.; Durieux, M.E. Arterial waveform analysis for the anesthesiologist: Past, present, and future concepts. *Anesth. Analg.* **2011**, *113*, 766–776. [CrossRef]

49. Wasicek, P.J.; Teeter, W.A.; Yang, S.; Hu, P.; Gamble, W.B.; Galvagno, S.M.; Hoehn, M.R.; Scalea, T.M.; Morrison, J.J. Arterial waveform morphomics during hemorrhagic shock. *Eur. J. Trauma Emerg. Surg. Off. Publ. Eur. Trauma Soc.* **2019**. [CrossRef]

50. Van der Ster, B.; Westerhof, B.; Stock, W.; Van Lieshout, J. Detecting central hypovolemia in simulated hypovolemic shock by automated feature extraction with principal component analysis. *Physiol. Rep.* **2018**, *6*, e13895. [CrossRef]

51. Holder, A.L.; Clermont, G. Using what you get: Dynamic physiologic signatures of critical illness. *Crit. Care Clin.* **2015**, *31*, 133–164. [CrossRef] [PubMed]

52. Esper, S.A.; Pinsky, M.R. Arterial waveform analysis. *Best Pract. Res. Clin. Anaesthesiol.* **2014**, *28*, 363–380. [CrossRef] [PubMed]

53. Reisner, A.; Shaltis, P.A.; McCombie, D.; Asada, H.H. Utility of the photoplethysmogram in circulatory monitoring. *Anesthesiology* **2008**, *108*, 950–958. [CrossRef] [PubMed]

54. Schiller, A.M.; Howard, J.T.; Lye, K.R.; Magby, C.G.; Convertino, V.A. Comparisons of traditional metabolic markers and compensatory reserve as early predictors of tolerance to central hypovolemia in humans. *Shock* **2018**, *50*, 71–77. [CrossRef] [PubMed]

55. Convertino, V.A.; Sawka, M.N. Wearable compensatory reserve measurement for hypovolemia sensing. *J. Appl. Physiol.* **2018**, *124*, 442–451. [CrossRef]

56. Schiller, A.M.; Howard, J.T.; Convertino, V.A. The physiology of blood loss and shock: New insights from a human laboratory model of hemorrhage. *Exp. Biol. Med.* **2017**, *242*, 874–883. [CrossRef] [PubMed]

57. Schlotman, T.E.; Akers, K.S.; Nessen, S.C.; Convertino, V.A. Differentiating compensatory mechanisms associated with low tolerance to central hypovolemia in women. *Am. J. Physiol. Heart Circ. Physiol.* **2019**, *316*, H609–H616. [CrossRef]

58. Wenner, M.M.; Hinds, K.; Howard, J.; Nawn, C.D.; Carter, R., III; Hinojosa-Laborde, C.; Stachenfeld, N.S.; Convertino, V. Differences in compensatory response to progressive reductions in central blood volume of African American and Caucasian women. *J. Trauma Acute Care Surg.* **2018**, *85*, S77–S83. [CrossRef]

59. Wenner, M.M.; Hinds, K.A.; Howard, J.T.; Nawn, C.D.; Stachenfeld, N.S.; Convertino, V.A. Measurement of compensatory reserve predicts racial differences in tolerance to simulated hemorrhage in women. *J. Trauma Acute Care Surg.* **2018**, *85*, S77–S83. [CrossRef]

60. Johnson, M.; Alarhayem, A.; Convertino, V.; Carter, R., 3rd; Chung, K.; Stewart, R.; Myers, J.; Dent, D.; Liao, L.; Cestero, R.; et al. Compensatory reserve index: Performance of a novel monitoring technology to identify the bleeding trauma patient. *Shock* **2018**, *49*, 295–300. [CrossRef]

61. Stewart, C.L.; Mulligan, J.; Grudic, G.Z.; Convertino, V.A.; Moulton, S.L. Detection of low-volume blood loss: The compensatory reserve index versus traditional vital signs. *J. Trauma Acute Care Surg.* **2014**, *77*, 892–897. [CrossRef]

62. Schisterman, E.F.; Perkins, N.J.; Liu, A.; Bondell, H. Optimal cut-point and its corresponding Youden Index to discriminate individuals using pooled blood samples. *Epidemiology* **2005**, *16*, 73–81. [CrossRef] [PubMed]

63. Schlotman, T.E.; Suresh, M.; Koons, N.J.; Howard, J.T.; Convertino, V.A. Comparisons of measures of compensatory reserve and heart rate variability as early indicators of hemodynamic decompensation in progressive hypovolemia. *J. Trauma Acute Care Surg.* **2020**, *89*, S161–S168. [PubMed]

64. Nadler, R.; Convertino, V.A.; Gendler, S.; Lending, G.; Lipsky, A.M.; Cardin, S.; Lowenthal, A.; Glassberg, E. The value of non-invasive mesurement of the compensatory reserve index in monitoring and triage of patients experiencing minimal blood loss. *Shock* **2014**, *42*, 93–98. [CrossRef] [PubMed]

65. Moulton, S.L.; Mulligan, J.; Santoro, M.A.; Bui, K.; Grudic, G.Z.; MacLeod, D. Validation of a noninvasive monitor to continuously trend individual responses to hypovolemia. *J. Trauma Acute Care Surg.* **2017**, *83*, S104–S111. [CrossRef] [PubMed]

66. Koons, N.J.; Owens, G.A.; Parsons, D.L.; Schauer, S.G.; Buller, J.L.; Convertino, V.A. Compensatory reserve: A novel monitoring capability for early detection of hemorrhage by combat medics. *J. Trauma Acute Care Surg.* **2020**, *89*, S146–S152.

67. Muniz, G.W.; Wampler, D.A.; Manifold, C.A.; Grudic, G.Z.; Mulligan, J.; Moulton, S.; Gerhardt, R.T.; Convertino, V.A. Promoting early diagnosis of hemodynamic instability during simulated hemorrhage with the use of a real-time decision-assist algorithm. *J. Trauma Acute Care Surg.* **2013**, *75*, S184–S189. [CrossRef]

68. Benov, A.; Yaslowitz, O.; Hakim, T.; Amir-Keret, R.; Nadler, R.; Brand, A.; Glassberg, E.; Yitzhak, A.; Convertino, V.A.; Paran, H. The effect of blood transfusion on compensatory reserve: A prospective clinical trial. *J. Trauma Acute Care Surg.* **2017**, *83*, S71–S76. [CrossRef]

69. Benov, A.; Brand, A.; Rozenblat, T.; Antebi, B.; Ben-Ari, A.; Amir-Keret, R.; Nadler, R.; Chen, J.; Chung, K.K.; Convertino, V.A.; et al. Evaluation of sepsis using compensatory reserve measurement: A prospective clinical trial. *J. Trauma Acute Care Surg.* **2020**, *89*, S153–S160. [CrossRef]

70. Stewart, C.L.; Nawn, C.D.; Mulligan, J.; Grudic, G.; Moulton, S.L.; Convertino, V.A. The Compensatory Reserve for Early and Accurate Prediction of Hemodynamic Compromise: Case Studies for Clinical Utility in Acute Care and Physical Performance. *J. Spec. Oper. Med.* **2016**, *16*, 6–13.

71. Stewart, C.L.; Mulligan, J.; Grudic, G.Z.; Talley, M.E.; Jurkovich, G.J.; Moulton, S.L. The Compensatory reserve index following injury: Results of a prospective clinical trial. *Shock* **2016**, *46*, 61–67. [CrossRef] [PubMed]

72. Stewart, C.L.; Mulligan, J.; Grudic, G.Z.; Pyle, L.; Moulton, S.L. A noninvasive computational method for fluid resuscitation monitoring in pediatric burns: A preliminary report. *J. Burn Care Res.* **2015**, *36*, 145–150. [CrossRef] [PubMed]

73. Imholz, B.P.M.; Wieling, W.; van Montfrans, G.A.; Wesseling, K.H. Fifteen years experience with finger arterial pressure monitoring. *Cardiovasc. Res.* **1998**, *38*, 605–616. [CrossRef]

74. Parati, G.; Casadei, R.; Groppelli, A.; Di Rienzo, M.; Mancia, G. Comparison of finger and intra-arterial blood pressure monitoring at rest and during laboratory testing. *Hypertension* **1989**, *13*, 647–655. [CrossRef]

75. Lee, J.; Nam, K.C. *Tonometric Vascular Function Assessment*; INTECH Open Access Publisher: London, UK, 2009.

76. Wang, Z.; Xu, Y. Design and optimization of an ultra-sensitive piezoresistive accelerometer for continuous respiratory sound monitoring. *Sens. Lett.* **2007**, *5*, 450–458. [CrossRef]

77. Shelley, K.H. Photoplethysmography: Beyond the calculation of arterial oxygen saturation and heart rate. *Anesth. Analg.* **2007**, *105*, S31–S36. [CrossRef]

78. Li, K.; Warren, S. A wireless reflectance pulse oximeter with digital baseline control for unfiltered photoplethysmograms. *IEEE Trans. Biomed. Circuits Syst.* **2012**, *6*, 269–278. [CrossRef]

79. Maeda, Y.; Sekine, M.; Tamura, T. Relationship between measurement site and motion artifacts in wearable reflected photoplethysmography. *J. Med. Syst.* **2011**, *35*, 969–976. [CrossRef]

80. Asada, H.H.; Shaltis, P.; Reisner, A.; Rhee, S.; Hutchinson, R.C. Mobile monitoring with wearable photoplethysmographic biosensors. *IEEE Eng. Med. Biol. Mag.* **2003**, *22*, 28–40. [CrossRef]

81. Chung, H.U.; Rwei, A.Y.; Hourlier-Fargette, A.; Xu, S.; Lee, K.; Dunne, E.C.; Xie, Z.; Liu, C.; Carlini, A.; Kim, D.H.; et al. Skin-interfaced biosensors for advanced wireless physiological monitoring in neonatal and pediatric intensive-care units. *Nat. Med.* **2020**, *26*, 418–429. [CrossRef]

82. Jeong, H.; Wang, L.; Ha, T.; Mitbander, R.; Yang, X.; Dai, Z.; Qiao, S.; Shen, L.; Sun, N.; Lu, N. Modular and reconfigurable wireless E-tattoos for personalized sensing. *Adv. Mater. Technol.* **2019**, *4*. [CrossRef]

83. Biswas, S.; Shao, Y.; Hachisu, T.; Nguyen-Dang, T.; Visell, Y. Integrated soft optoelectronics for wearable health monitoring. *Adv. Mater. Technol.* **2020**, *5*, 2000347. [CrossRef]

84. Han, D.; Khan, Y.; Ting, J.; Zhu, J.; Combe, C.; Wadsworth, A.; McCulloch, I.; Arias, A.C. Pulse oximetry using organic optoelectronics under ambient light. *Adv. Mater. Technol.* **2020**, *5*, 1901122. [CrossRef]

85. Drzewiecki, G.M.; Melbin, J.; Noordergraaf, A. *Deformational Forces in Arterial Tonometry*; IEEE: New York, NY, USA, 1984; Volume 28.

86. Eckerle, J.S. Tonometry, arterial. In *Encyclopedia of Medical Devices and Instrumentation*; Webster, J.G., Ed.; John Wiley & Sons: New York, NY, USA, 1988.

87. Schwartz, G.; Tee, B.C.-K.; Mei, J.; Appleton, A.L.; Kim, D.H.; Wang, H.; Bao, Z. Flexible polymer transistors with high pressure sensitivity for application in electronic skin and health monitoring. *Nat. Commun.* **2013**, *4*, 1859. [CrossRef]

88. Wang, C.; Li, X.; Hu, H.; Zhang, L.; Huang, Z.; Lin, M.; Zhang, Z.; Yin, Z.; Huang, B.; Gong, H. Monitoring of the central blood pressure waveform via a conformal ultrasonic device. *Nat. Biomed. Eng.* **2018**, *2*, 687–695. [CrossRef] [PubMed]

89. Tavakolian, K.; Dumont, G.A.; Houlton, G.; Blaber, A.P. Precordial vibrations provide noninvasive detection of early-stage hemorrhage. *Shock* **2014**, *41*, 91–96. [CrossRef] [PubMed]

90. Zia, J.; Kimball, J.; Rolfes, C.; Hahn, J.-O.; Inan, O.T. Enabling the assessment of trauma-induced hemorrhage via smart wearable systems. *Sci. Adv.* **2020**, *6*, eabb1708. [CrossRef] [PubMed]

91. Inan, O.T.; Migeotte, P.F.; Kwang-Suk, P.; Etemadi, M.; Tavakolian, K.; Casanella, R.; Zanetti, J.; Tank, J.; Funtova, I.; Prisk, G.K.; et al. Ballistocardiography and seismocardiography: A review of recent advances. *IEEE J. Biomed. Health Inform.* **2015**, *19*, 1414–1427. [CrossRef]

92. Etemadi, M.; Inan, O.T. Wearable ballistocardiogram and seismocardiogram systems for health and performance. *J. Appl. Physiol.* **2017**, *124*, 452–461. [CrossRef]

93. Di Rienzo, M.; Rizzo, G.; Işılay, Z.M.; Lombardi, P. SeisMote: A Multi-Sensor Wireless platform for cardiovascular monitoring in laboratory, daily life, and telemedicine. *Sensors* **2020**, *20*, 680. [CrossRef]

94. Inan, O.T.; Baran Pouyan, M.; Javaid, A.Q.; Dowling, S.; Etemadi, M.; Dorier, A.; Heller, J.A.; Bicen, A.O.; Roy, S.; De Marco, T.; et al. Novel Wearable Seismocardiography and Machine Learning Algorithms Can Assess Clinical Status of Heart Failure Patients. *Circ. Heart Fail.* **2018**, *11*, e004313. [CrossRef] [PubMed]

95. Shandhi, M.M.H.; Hersek, S.; Fan, J.; Sander, E.; De Marco, T.; Heller, J.A.; Etemadi, M.; Klein, L.; Inan, O.T. Wearable patch based estimation of oxygen uptake and assessment of clinical status during cardiopulmonary exercise testing in patients with heart failure. *J. Card. Fail.* **2020**, in press. [CrossRef] [PubMed]

96. Liu, Y.; Norton, J.J.; Qazi, R.; Zou, Z.; Ammann, K.R.; Liu, H.; Yan, L.; Tran, P.L.; Jang, K.I.; Lee, J.W.; et al. Epidermal mechano-acoustic sensing electronics for cardiovascular diagnostics and human-machine interfaces. *Sci. Adv.* **2016**, *2*, e1601185. [CrossRef] [PubMed]

97. Boutry, C.M.; Nguyen, A.; Lawal, Q.O.; Chortos, A.; Rondeau-Gagne, S.; Bao, Z. A sensitive and biodegradable pressure sensor array for cardiovascular monitoring. *Adv. Mater.* **2015**, *27*, 6954–6961. [CrossRef] [PubMed]

98. Imani, S.; Bandodkar, A.J.; Vinu Mohan, A.M.; Kumar, R.; Yu, S.; Wang, J.; Mercier, P.P. A wearable chemical–electrophysiological hybrid biosensing system for real-time health and fitness monitoring. *Nat. Commun.* **2016**, *7*, 1–7. [CrossRef] [PubMed]

99. Sel, K.; Ibrahim, B.; Jafari, R. ImpediBands: Body coupled bio-impedance patches for physiological sensing proof of concept. *IEEE Trans. Biomed. Circuits Syst.* **2020**, *14*, 757–774. [CrossRef]

100. Teichmann, D.; Kuhn, A.; Leonhardt, S.; Walter, M. The MAIN Shirt: A textile-integrated magnetic induction sensor array. *Sensors* **2014**, *14*, 1039–1056. [CrossRef]

101. Ganti, V.G.; Carek, A.; Nevius, B.N.; Heller, J.; Etemadi, M.; Inan, O.T. Wearable cuff-less blood pressure estimation at home via pulse transit time. *IEEE J. Biomed. Health Inform.* **2020**, in press. [CrossRef]

102. Lee, J.; Matsumura, K.; Yamakoshi, K.-i.; Rolfe, P.; Tanaka, S.; Yamakoshi, T. Comparison between Red, Green and Blue Light Reflection Photoplethysmography for Heart Rate Monitoring During Motion. In Proceedings of the Engineering in Medicine and Biology Society (EMBC), Osaka, Japan, 3–7 July 2013; pp. 1724–1727.

103. Spigulis, J.; Gailite, L.; Lihachev, A.; Erts, R. Simultaneous recording of skin blood pulsations at different vascular depths by multiwavelength photoplethysmography. *Appl. Opt.* **2007**, *46*, 1754–1759. [CrossRef]

104. Tamura, T.; Maeda, Y.; Sekine, M.; Yoshida, M. Wearable photoplethysmographic sensors—past and present. *Electronics* **2014**, *3*, 282–302. [CrossRef]

105. Kaisti, M.; Panula, T.; Leppanen, J.; Punkkinen, R.; Jafari Tadi, M.; Vasankari, T.; Jaakkola, S.; Kiviniemi, T.; Airaksinen, J.; Kostiainen, P.; et al. Clinical assessment of a non-invasive wearable MEMS pressure sensor array for monitoring of arterial pulse waveform, heart rate and detection of atrial fibrillation. *NPJ Digit. Med.* **2019**, *2*, 39. [CrossRef] [PubMed]

106. Digiglio, P.; Li, R.; Wang, W.; Pan, T. Microflotronic arterial tonometry for continuous wearable non-invasive hemodynamic monitoring. *Ann. Biomed. Eng.* **2014**, *42*, 2278–2288. [CrossRef] [PubMed]

107. Hossein, A.; Mirica, D.C.; Rabineau, J.; Rio, J.I.D.; Morra, S.; Gorlier, D.; Nonclercq, A.; van de Borne, P.; Migeotte, P.F. Accurate detection of dobutamine-induced haemodynamic changes by kino-cardiography: A randomised double-blind placebo-controlled validation study. *Sci. Rep.* **2019**, *9*, 10479. [CrossRef] [PubMed]

108. Poh, M.-Z.; Swenson, N.C.; Picard, R.W. Motion-tolerant magnetic earring sensor and wireless earpiece for wearable photoplethysmography. *IEEE Trans. Inf. Technol. Biomed.* **2010**, *14*, 786–794. [CrossRef] [PubMed]

109. Han, H.; Kim, M.-J.; Kim, J. Development of Real-Time Motion Artifact Reduction Algorithm for a Wearable Photoplethysmography. In Proceedings of the Engineering in Medicine and Biology Society, Lyon, France, 23–26 August 2007; pp. 1538–1541.

110. Wood, L.B.; Asada, H.H. Noise Cancellation Model Validation for Reduced Motion Artifact Wearable PPG Sensors Using MEMS Accelerometers. In Proceedings of the 2006 International Conference of the IEEE Engineering in Medicine and Biology Society, New York, NY, USA, 30 August–3 September 2006; pp. 3525–3528.

111. Javaid, A.Q.; Ashouri, H.; Dorier, A.; Etemadi, M.; Heller, J.A.; Roy, S.; Inan, O.T. Quantifying and reducing Motion artifacts in wearable seismocardiogram measurements during walking to assess left ventricular health. *IEEE Trans. Biomed. Eng.* **2017**, *64*, 1277–1286. [CrossRef]

112. Yang, C.; Tavassolian, N. Motion artifact cancellation of seismocardiographic recording from moving subjects. *IEEE Sens. J.* **2016**, *16*, 5702–5708. [CrossRef]

113. Inan, O.T.; Dookun, P.; Giovangrandi, L.; Kovacs, G.T. Noninvasive measurement of physiological signals on a modified home bathroom scale. *IEEE Trans. Biomed. Eng.* **2012**, *59*, 2137–2143. [CrossRef]

114. Inan, O.T.; Etemadi, M.; Widrow, B.; Kovacs, G.T.A. Adaptive cancellation of floor vibrations in standing ballistocardiogram measurements using a seismic sensor as a noise reference. *IEEE Trans. Biomed. Eng.* **2010**, *57*, 722–727. [CrossRef]

115. Orphanidou, C.; Bonnici, T.; Charlton, P.; Clifton, D.; Vallance, D.; Tarassenko, L. Signal-quality indices for the electrocardiogram and photoplethysmogram: Derivation and applications to wireless monitoring. *IEEE J. Biomed. Health Inform.* **2014**, *19*, 832–838. [CrossRef]

116. Li, Q.; Clifford, G.D. Dynamic time warping and machine learning for signal quality assessment of pulsatile signals. *Physiol. Meas.* **2012**, *33*, 1491–1501. [CrossRef]

117. Zia, J.S.; Kimball, J.; Hersek, S.; Shandhi, M.; Semiz, B.; Inan, O. A unified framework for quality indexing and classification of seismocardiogram signals. *IEEE J. Biomed. Health Inform.* **2019**. [CrossRef] [PubMed]

118. Keogh, E.; Ratanamahatana, C.A. Exact indexing of dynamic time warping. *Knowl. Inf. Syst.* **2005**, *7*, 358–386. [CrossRef]

119. Zia, J.S.; Kimball, J.; Rozell, C.J.; Inan, O. Harnessing the manifold structure of cardiomechanical signals for physiological monitoring during hemorrhage. *IEEE Trans. Biomed. Eng.* **2020**. [CrossRef]

120. Karmer, G.C.; Kinsky, M.P.; Prough, D.S.; Salinas, J.; Sondeen, J.L.; Hazel-Scero, M.L.; Mitchell, C.E. Closed-loop control of fluid therapy for treatment of hypovolemia. *J. Trauma* **2008**, *64*, S333–S341. [CrossRef] [PubMed]

121. Libert, N.; Chenegros, G.; Harrois, A.; Baudry, N.; Cordurie, G.; Benosman, R.; Vicaut, E.; Duranteau, J. Performance of closed-loop resuscitation of haemorrhagic shock with fluid alone or in combination with norepinephrine: An experimental study. *Ann. Intensive Care* **2018**, *8*, 89. [CrossRef] [PubMed]

122. Bighamian, R.; Kim, C.-S.; Reisner, A.T.; Hahn, J.-O. Closed-loop fluid resuscitation control via blood volume estimation. *J. Dyn. Syst. Meas. Control* **2016**, *138*, 111005. [CrossRef]

123. Schauer, S.G.; Naylor, J.F.; Oliver, J.J.; Maddry, J.K.; April, M.D. An analysis of casualties presenting to military emergency departments in Iraq and Afghanistan. *Am. J. Emerg. Med.* **2019**, *37*, 94–99. [CrossRef]

124. Chen, F.; Chandrakasan, A.P.; Stojanovic, V.M. Design and analysis of a hardware-efficient compressed sensing architecture for data compression in wireless sensors. *IEEE J. Solid State Circuits* **2012**, *47*, 744–756. [CrossRef]

125. Craven, D.; McGinley, B.; Kilmartin, L.; Glavin, M.; Jones, E. Compressed sensing for bioelectric signals: A review. *IEEE J. Biomed. Health Inform.* **2015**, *19*, 529–540. [CrossRef]

126. Butler, F.K.; Holcomb, J.B.; Schreiber, M.A.; Kotwal, R.S.; Jenkins, D.A.; Champion, H.R.; Bowling, F.; Cap, A.P.; Dubose, J.J.; Dorlac, W.C.; et al. Fluid resuscitation for hemorrhagic shock in tactical combat casualty care: TCCC Guidelines Change 14-01-2 June 2014. *J. Spec. Oper. Med.* **2014**, *14*, 30–55.
127. Mulder, M.B.; Eidelson, S.A.; Buzzelli, M.D.; Gross, K.R.; Batchinsky, A.I.; Convertino, V.A.; Schulman, C.I.; Namias, N.; Proctor, K.G. Exercise-induced changes in compensatory reserve and heart rate complexity. *Aerosp. Med. Hum. Perform.* **2019**, *90*, 1009–1015. [CrossRef] [PubMed]

Publisher's Note: MDPI stays neutral with regard to jurisdictional claims in published maps and institutional affiliations.

Article

Vital-Signs Detector Based on Frequency-Shift Keying Radar

Jae Young Sim, Jae-Hyun Park and Jong-Ryul Yang * 

Department of Electronic Engineering, Yeungnam University, Gyeongsan, Gyeongbuk 38541, Korea;
ja2922@yu.ac.kr (J.Y.S.); bravopark@ynu.ac.kr (J.-H.P.)
* Correspondence: jryang@yu.ac.kr; Tel.: +82-53-810-2495

Received: 9 September 2020; Accepted: 24 September 2020; Published: 26 September 2020

Abstract: A frequency-shift keying (FSK) radar in the 2.45-GHz band is proposed for highly accurate vital-signs detection. The measurement accuracy of the proposed detector for the heartbeat is increased by using the cross-correlation between the phase differences of signals at two frequencies used by the FSK radar, which alternately transmits and receives the signals with different frequencies. Two frequencies—2.45 and 2.5 GHz—are effectively discriminated by using the envelope detection with the frequency control signal of the signal generator in the output waveform of the FSK radar. The phase difference between transmitted and received signals at each frequency is determined after calibrating the I/Q imbalance and direct-current offset using a data-based imbalance compensation algorithm, the Gram–Schmidt procedure, and the Pratt method. The absolute-distance measurement results for a human being show that the vital signs obtained at each frequency using the proposed FSK radar have a cross-correlation. The heartbeat detection results for the proposed FSK radar at a distance of < 2.4 m indicate a reduction in the error rate and an increase in the signal-to-noise ratio compared with those obtained using a single operating frequency.

Keywords: frequency-shift keying radar; cross-correlation; envelope detection; continuous-wave radar; frequency discrimination; vital-signs monitoring; heartbeat accuracy improvement; heartbeat detection; absolute distance measurement; radar signal processing

---

1. Introduction

Research on vital-signs detection using radar technology has been performed since the short-range radar system was introduced for the detection of human vital signs in the 1970s [1]. A vital-signs detector based on radar technology can measure the vital signs without electrode contacts or restriction of the measurement environment, in contrast to contact-type detectors [2]. Owing to these advantages, the radar sensor for vital-signs detection is promising as a key element in continuous monitoring systems for home-care service, local positioning and tracking in disaster scenes such as earthquakes and fires, as well as disease identification using heartrate variability (HRV) analysis, e.g., sleep apnea and angina pectoris [3–5]. In particular, the study on a vital-signs detector using radar technology aims to achieve a level of detection accuracy so as to fully replace the electrocardiogram (ECG) sensor in medical applications, such as HRV analysis.

Among the radar technologies, continuous-wave (CW) Doppler radars are useful for vital-signs monitoring based on periodic motions in the human body because of their simple hardware configuration and signal processing [3–6]. However, CW radars have a limitation in that noise can be increased or the receiver can be saturated by the movement of the target or surrounding clutters, as signals caused by all movements are collected by the antenna [7]. This limitation causes particularly significant degradation of accuracy in heartbeat detection compared with respiration detection, because the chest movement caused by the heartbeat is 0.2–0.5 mm, whereas the chest movement caused by respiration is 4–12 mm [8,9]. It is important to implement the CW radar with a high signal-to-noise ratio (SNR) for heartbeat detection to increase the detection accuracy [10]. The SNR in the CW radar can be improved by increasing the transmitted power, but the maximum allowable effective isotropic radiated power is restricted by the regulation in each frequency band. It is not easy to implement a radar front-end with a high SNR by using a low-noise and high-gain design methodology, and a complex radar architecture including calibration circuits and a calibration process could be needed to improve the SNR [10]. The accuracy of vital-signs detection can be also improved by using signal-processing techniques such as autocorrelation, the wavelet transform, and cross-correlation [11–14]. The autocorrelation method improves the detection accuracy for the periodic signal by converging signals to the most representative period frequency but has limitations for accurately monitoring heartbeat signals that change over time and evaluating their variability [11,14]. The wavelet transform is useful for increasing the accuracy by effectively extracting peaks of the heartbeat signal, but it is difficult to develop a generalized wavelet function that can improve the accuracy while being independent of the measurement environment, such as the characteristics of the subject and the clutter in the surroundings [12,14]. The cross-correlation method was proposed for increasing the detection accuracy by exploiting the similarity between vital signs independently measured at multiple frequencies by using a dual-band antenna and commercialized measurement equipment [13]. Although this approach is useful for achieving a high accuracy because the vital signs are independent of the characteristics of the radar and the operation frequency, it is necessary to use several radars with different operating frequencies, and customized components such as a dual-band antenna are required.

In this work, a vital-signs detector with improved accuracy based on frequency-shift keying (FSK) radar technology is proposed. An FSK radar is used for a highly accurate range detection based on the phase difference between the transmitting and receiving signals separately obtained at more than two operating frequencies [15]. The vital-signs detection in the FSK radar is performed by using the same method as the CW Doppler radar as the FSK radar can be regarded as operating with several CW signals in the same hardware configuration. The proposed FSK radar improves the detection accuracy and SNR by using the cross-correlation between the vital signs independently obtained from each operating frequency. A method of effectively discriminating two frequency signals at the baseband output is necessary for accurately obtaining the results of the cross-correlation in the FSK radar [16]. This work presents the signal discrimination technique based on envelope detection in synchronization with a frequency control signal as a method for separately obtaining each operating frequency signal. The proposed technique can discriminate the phase information at each operating frequency from the FSK radar in a short period, and the cross-correlation for improving the detection accuracy can be easily implemented in the single radar by using this technique. The imbalance between the in-phase (I) and quadrature (Q) signals and a direct-current (DC) offset, which are critical characteristics for the accuracy of the phase measurement, are calibrated by modifying the method used by the CW Doppler radar sensor for the FSK radar. Distance measurements with the proposed radar indicate that the phase difference of vital signs at each operating frequency have a cross-correlation. The measurement results of heartbeat detection have high accuracy relative to the reference electrocardiogram (ECG) signal owing to the cross-correlation of the proposed FSK radar. The measurement results for the distance and heartbeat show that the detection accuracy at each frequency depends on the distance, owing to the characteristics of the FSK radar. The operating principles of the FSK radar for vital-signs detection are described in Section 2. Section 3 shows the implementation method of the proposed FSK radar, including the calibration process and digital signal processing. The measurement results for vital signs and distances and an analysis of the results discussed in Section 4. Conclusions are presented in Section 5.

2. Vital-Signs Detection Using FSK Radar

2.1. Operating Principles for FSK Radar

An FSK radar measures the information of the target from the phase differences between two operating frequencies. The phase difference is obtained by comparing the transmitted and received signals at a single CW operating frequency. Accurate information can be obtained by increasing the measurement accuracy of the phase difference at each frequency. A block diagram of the FSK radar is shown in Figure 1. A quadrature architecture is adopted in the proposed FSK radar to avoid the null point problem in the CW Doppler radar [17,18]. Two discrete frequencies f_1 and f_2 are alternately transmitted for the switching time period of T with a duty cycle of 50% by a signal generator. A short time period is advantageous for increasing the measurement accuracy because the FSK radar assumes that the information of the target is constant during the period. However, a sufficient period for sampling baseband signals is needed to obtain the phase difference at each frequency. Thus, the time period should be optimized with consideration of the maximum velocity of the movement in the target and the maximum sampling frequency of the synchronous data-acquisition (DAQ) device.

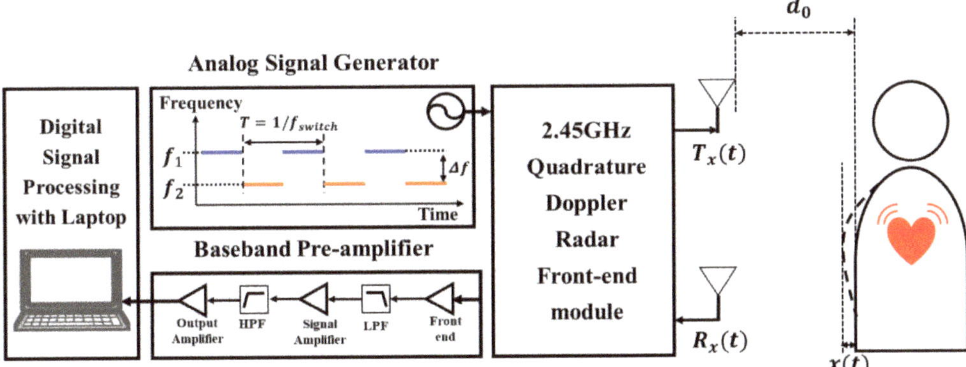

Figure 1. Block diagram of the proposed frequency-shift keying (FSK) radar.

The transmitted signals $T_x(t)$ in FSK radar can be expressed as follows:

$$T_x(t) = \begin{cases} A_{T1} \cdot \cos[2\pi f_1 t + \varphi_1(t)], & 0 < t \le \frac{T}{2} \\ A_{T2} \cdot \cos[2\pi f_2 t + \varphi_2(t)], & \frac{T}{2} < t \le T \end{cases} \tag{1}$$

where A_{T1} and A_{T2} represent the amplitudes of the transmitted signals, and $\varphi_1(t)$ and $\varphi_2(t)$ represent the phase noises of the two transmitted frequencies in the signal generator, respectively. The receiving signals in the radar are modulated to the Doppler frequencies produced by the chest movements caused by the respiration and heartbeat [18]. The received signals $R_x(t)$ in FSK radar can be expressed as follows:

$$R_x(t) \approx \begin{cases} A_{R1} \cdot \cos\left[2\pi f_1 t - \frac{4\pi d_0}{\lambda_1} - \frac{4\pi x_1(t)}{\lambda_1} + \varphi_1\left(t - \frac{2d_0}{c}\right)\right], & 0 < t \le \frac{T}{2} \\ A_{R2} \cdot \cos\left[2\pi f_2 t - \frac{4\pi d_0}{\lambda_2} - \frac{4\pi x_2(t)}{\lambda_2} + \varphi_2\left(t - \frac{2d_0}{c}\right)\right], & \frac{T}{2} < t \le T \end{cases} \tag{2}$$

where A_{R1} and A_{R2} represent the amplitudes of the received signals; c represents the velocity of light, λ_1 and λ_2 represent the wavelengths of the two frequencies, respectively; d_0 represents the fixed distance between the radar and the target; and $x_1(t)$ and $x_2(t)$ represents the displacements of the chest caused by the respiration and the heartbeat. The phase noise of the received signal is described with the time delay of $2d_0/c$ by considering the time of the round trip of the signal at the distance. The in-phase (I_1

and I_2) and quadrature (Q_1 and Q_2) baseband signals, which are obtained from the down-conversion quadrature mixers, can be expressed as follows:

$$I_k(t) = A_I \cdot \cos\left[\frac{4\pi d_0}{\lambda_k} + \frac{4\pi x_k(t)}{\lambda_k} + \Delta\varphi_k(t)\right] + DCI_k, \; k = 1, 2, \text{ and} \tag{3}$$

$$Q_k(t) = A_I A_E \cdot \sin\left[\frac{4\pi d_0}{\lambda_k} + \frac{4\pi x_k(t)}{\lambda_k} + \Delta\varphi_k(t) + \phi_E\right] + DCQ_k, \; k = 1, 2, \tag{4}$$

where A_I represents the amplitude of the I-channel baseband signal; A_E and ϕ_E represent the errors of the amplitude and phase, respectively; DCI_k and DCQ_k represent the DC offset voltages in the I and Q channels, respectively; and $\Delta\varphi_k$ represents the residual phase noise, which is the difference in the phase noise between the transmitted and received signals. The residual phase noise in the radar system can generally be neglected in the measurement of the distance and vital signs, owing to the range correlation effect [19]. The phase difference θ_k between the transmitted and received signals at each frequency can be represented by using (3) and (4) as follows:

$$\theta_k \cong \frac{4\pi}{\lambda_k}(d_0 + x_k(t)), \; k = 1, 2, \tag{5}$$

which is identical to the difference of the CW Doppler radar. The detectable range in the phase difference, which is 0–2π, is determined by the characteristics of the trigonometric function, as indicated by (5). The distance d_0 can be measured from the subtraction in each phase difference using the two frequencies in the FSK radar, as follows:

$$d_0 = \frac{c}{4\pi(f_1 - f_2)}(\theta_1 - \theta_2) - [x_1(t) - x_2(t)]. \tag{6}$$

If the difference in the vital signs generated during the transmitting and receiving signals of each frequency can be neglected, the absolute distance can be obtained as follows:

$$d_0 \cong \frac{c}{4\pi(f_1 - f_2)}(\theta_1 - \theta_2). \tag{7}$$

Thus, the error of the distance measurement in the FSK radar can show the cross-correlation of the vital signs at each operating frequency, and a low error corresponds to a high correlation rate between two vital signals. The periodic signals in the baseband can represent the vital signs included in the phase difference, because the FSK radar can be regarded as a CW Doppler radar with independent single-frequency operation. When the human motion does not have periodicity or is located outside of the frequency band of the vital signs, the vital signs can be detected through fast-Fourier transform (FFT) if the receiver is not saturated by the motion. The measured vital signs in the frequency band can be expressed as follows:

$$X(f) = \frac{\lambda_k}{4\pi} \int_{-\infty}^{\infty} \theta_k(t)e^{-j2\pi f_m t} dt. \tag{8}$$

2.2. Cross-Correlation Method

The cross-correlation method can be used for increasing the power of a periodic signal in a noisy environment [20]. Respiration and heartbeat, which are vital signs obtained directly from the CW radar, are both periodic signals, but the heartbeat of a stationary subject has a relatively low SNR compared with the respiration of the subject. In the proposed FSK radar, the cross-correlation method is used for accurately extracting the heartbeat from the raw data. The cross-correlation between the phase differences of signals at two operating frequencies can be mathematically expressed in the digital domain as follows:

$$\hat{R}_{\theta_1 \theta_2}[m] = \sum_{n=0}^{N-m-1} \theta_1[n+m]\theta_2^*[n], \ m = 1, 2, \ldots, 2N-1, \tag{9}$$

where $\hat{R}_{\theta_1 \theta_2}$ represents the result of the raw correlation between the nth element of θ_1 and the mth element of θ_2, and N represents the product of time and the sampling frequency [20]. The power levels of the elements simultaneously present in the two phase-different signals are increased using the cross-correlation method. The cross-correlated signal is expanded to double the length of the phase-difference signal in the time domain, as shown in Figure 2a. The frequency resolution of the signal is also increased in the frequency domain, as shown in Figure 2b, because the number of sampled data is increased at the same sampling frequency by the cross-correlation. Both white noise and noise signals caused by the hardware components that have different frequency characteristics are reduced by the cross-correlation. The proposed FSK radar can easily improve the vital-signs detection performance without additional signal processing, because the FSK radar uses the same hardware configuration to obtain the cross-correlated signal with the measured phase differences at the two operating frequencies. Thus, the peak detection accuracy of the heartbeat can be improved in the frequency domain because the SNR of the heartbeat in the proposed FSK radar is improved by the enhancement of the signal power level and the suppression of the noise level.

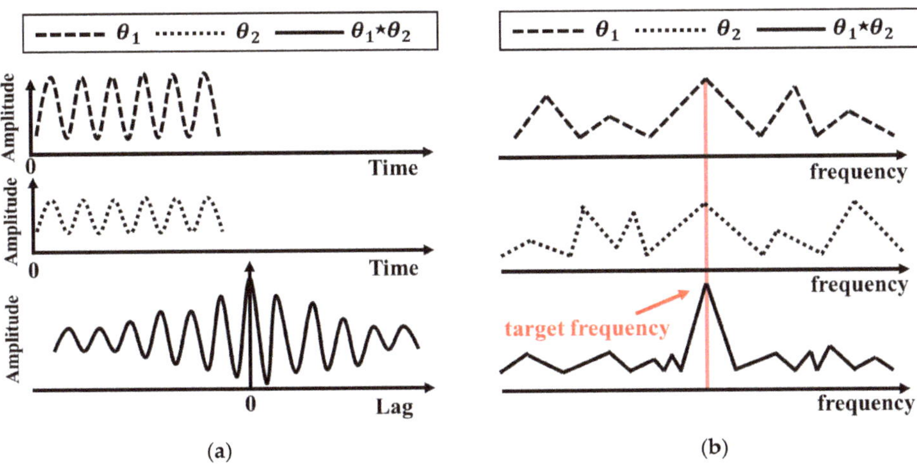

Figure 2. Cross-correlation method: (**a**) In the time domain; (**b**) in the frequency domain.

2.3. Frequency Discrimination Using Envelope Detection for Proposed FSK Radar

The discrimination of phase-difference signals at two different operating frequencies is the most important process for implementing the FSK radar. However, the signals at two frequencies are not continuous by the FSK operation, and the sampling points at each frequency are not identical. An envelope detection method can discriminate two nonoverlapped signals with different DC offsets at each frequency, because the offset characteristics of the radar components depend on the operating frequency [16]. This method is implemented using spline interpolation with not-a-knot conditions over local maxima separated by 10 samples. The baseband signals can be discriminated in each frequency by using the envelope detection method, but it cannot be known whether the discriminated signal shows anything of the two frequencies. The control signal for frequency switching in the signal generator is used to store each data-set of I/Q channel signals with each discriminated frequency in four data-sets. In the initial state, the stored data-set is regarded as the signals at f_1 when the control signal is high, and the data-set is regarded as the signals at f_2 when the control signal is low. The frequency is adjusted to ensure that the distance obtained from each data-set is positive. Figure 3a,b show the baseband *I/Q* signals for each frequency extracted from the raw data by the envelope detection method.

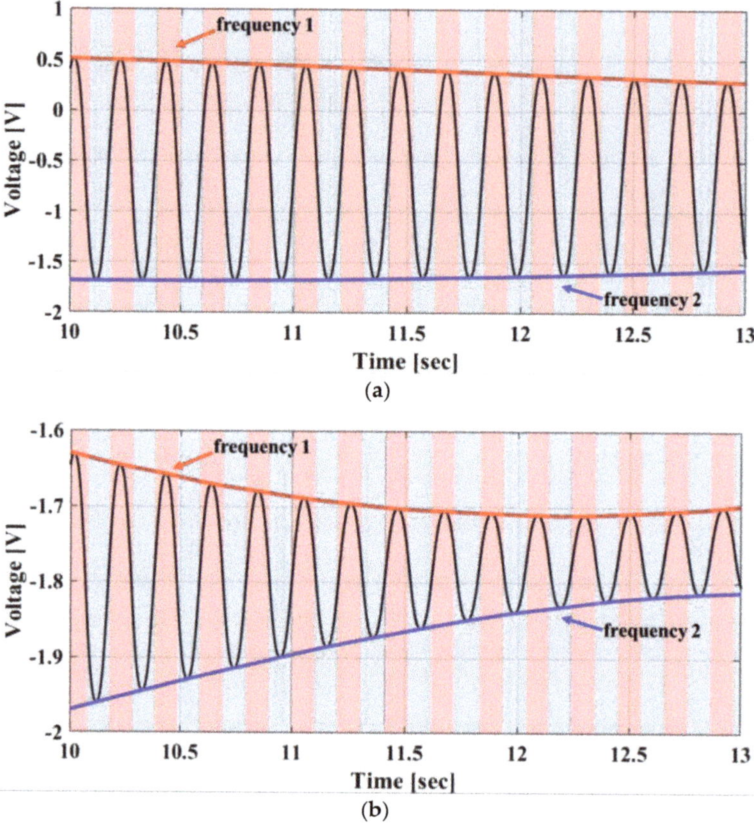

Figure 3. Proposed envelope detection method for discriminating the baseband signals at each operating frequency of the FSK radar. The frequency-control signal of the signal generator, which is indicated by red and blue shaders in the waveform, is used for synchronous data acquisition: (**a**) In-phase channel signals; (**b**) quadrature channel signals.

3. Implementation

3.1. Digital Signal Processing

The detection accuracies of both the vital signs of the subject and the distance to the subject are determined by how accurately the phase difference is measured in the FSK radar, and they are significantly affected by the *I/Q* imbalance and DC offset in the baseband. The *I/Q* imbalance in the amplitude and phase caused by the imperfections of the quadrature receiver can be measured by phase shifters and calibrated by the Gram–Schmidt procedure [21]. The phase shifter is used to generate a circle trajectory by varying the phases of the signal on the complex plane, and several other methods can be employed to implement this function. When the target is mechanically moved within a displacement similar to the half wavelength of the operating frequency, baseband signals can draw the circle trajectory on the complex plane as in the case of using the phase shifter [22]. However, an elliptical trajectory is generated owing to the change in the distance to the target, in contrast to the circle trajectory using phase shifters. The elliptical trajectory can be compensated by the data-based quadrature imbalance compensation technique using the ellipse fitting method [23]. Figure 4a,b present the calibrated *I/Q* channels after the imbalance calibration.

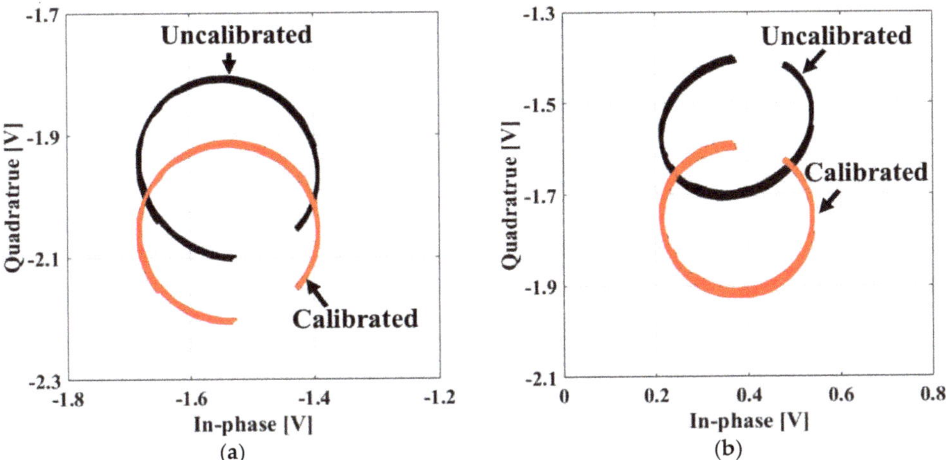

Figure 4. In-phase/quadrature (I/Q) imbalance calibration on the complex plane using the reference distance and the known periodic movement: (**a**) At the frequency of 2.45 GHz; (**b**) at the frequency of 2.5 GHz.

DC offsets are generated by stationary clutters in the surroundings as well as the imperfection of the hardware configuration of the radar. The FSK radar receives the vector-sum signals reflected from all objects located in the antenna beamwidth [15]. The accuracy of the phase difference can decrease as the measurement distance increases because the DC offsets increase with the amount of clutters received on the radar. The DC offsets on the radar can be eliminated while preserving vital signs, which are located near DC in the frequency domain, using a dynamic DC offset compensation algorithm [24]. Figure 5 presents the baseband *I/Q* signals at each frequency measured by the proposed FSK radar on the complex plane. The trajectory at each frequency is a part of each circle on the complex plane, and the DC offset voltage is indicated by the center of the circle by the circle fitting method using the trajectory. The DC offset can be effectively calibrated by this procedure, which is known as the Pratt method, when the circle trajectory is obtained from the measured data [25].

The phase difference in each frequency is extracted by demodulating the calibrated signals. In the proposed FSK radar, the arc-sine demodulation and the complex signal demodulation (CSD) techniques are used for detecting the vital signs of the subject and the distance to the subject, respectively [26,27]. The arc-sine demodulation technique is generally used as a demodulation technique, but is not suitable for distance measurement with the FSK radar, because the measurement error can significantly increase with the decreasing accuracy of the DC offset when accurate circle fitting cannot be realized via either the random body movement or the curved chest wall of the human body. The CSD technique is an appropriate demodulation method in distance measurement of the FSK radar, because an accurate circle fitting process is not mandatory in the CSD, in contrast to the arc-sine demodulation technique [28]. Bandpass filters with cutoff frequencies ranging from 0.8 to 2 Hz and from 0.1 to 0.8 Hz are used after the demodulation process to extract the respiration and heartbeat signals, respectively, in these two techniques. Considering the characteristics of the vital signs, which vary irregularly, FFT with a sliding window of 30 s is performed every 10 s with a measurement time of 90 s. Figure 6 shows the signal processing procedure to obtain the vital signs from the raw data of the proposed radar.

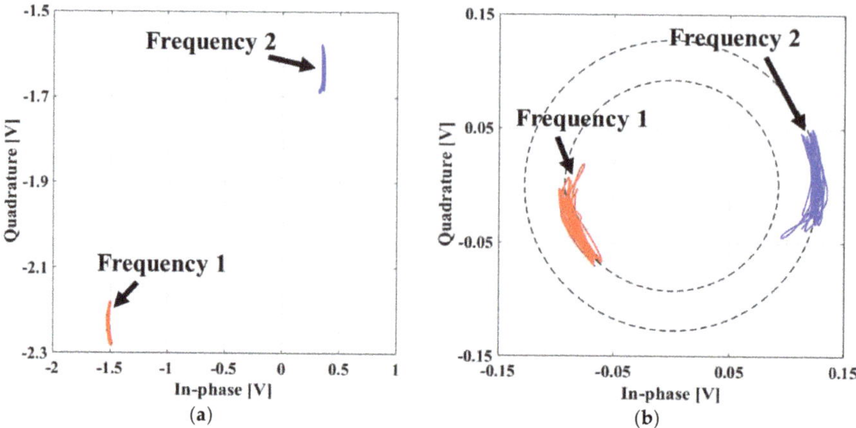

Figure 5. Direct-current (DC) offset calibration on the complex plane: (**a**) Uncalibrated signals of each frequency in the baseband; (**b**) Calibrated signals of each frequency based on the Pratt method.

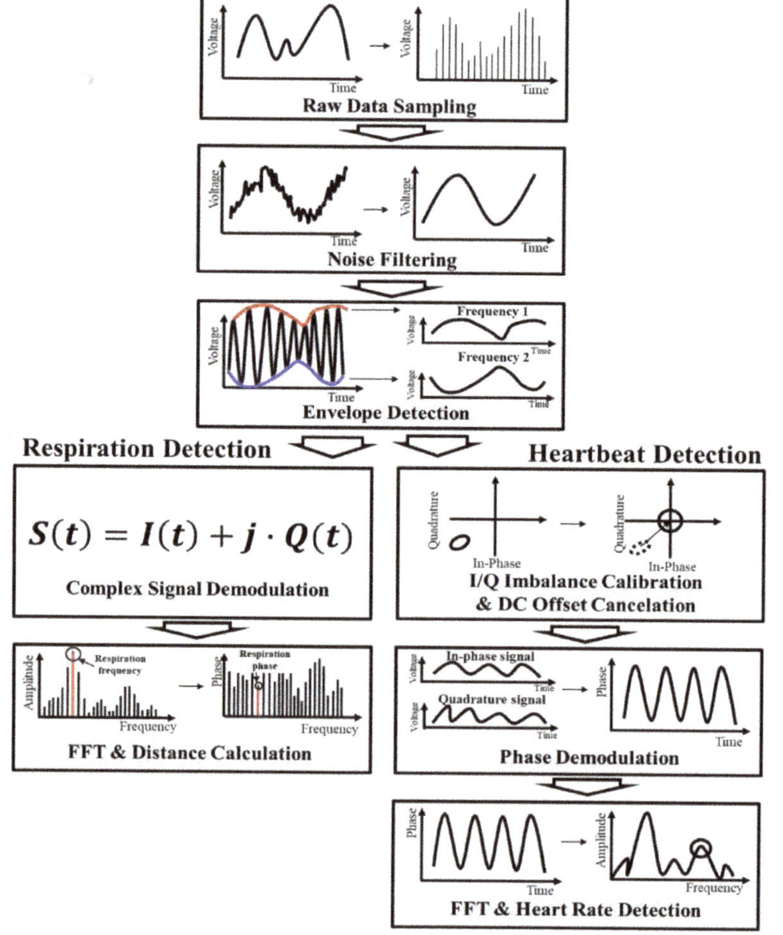

Figure 6. Signal processing procedure in the proposed vital-signs detector.

3.2. Implemented FSK Radar Module

For vital-signs and distance measurement, the frequencies of 2.45 and 2.5 GHz are used in the proposed FSK radar, which operates in the 2.45 GHz ISM band. The maximum unambiguous range of the FSK radar is 3 m, which is determined by the frequency difference of 50 MHz. The radar front-end circuit and two patch antennas are implemented on an FR4 printed circuit board (PCB) with a thickness of 1 mm, as shown in Figure 7. The FSK signals with a frequency spacing of 50 MHz and an output power of 15 dBm are generated by an N5183B signal generator manufactured by Keysight Technologies Inc. The FSK operation is realized by using the internal function of the signal generator. The generated signals are divided by a Wilkinson power divider into the reference and transmitting signals. Quadrature signals are generated by a hybrid power divider with a phase difference of 90° between two outputs. The signal of each frequency is radiated toward the subject and received by the separated patch antennas with a directivity of 5.9 dBi. The received signal is amplified using a low-noise amplifier (LNA) with a power gain of 13.7 dB and noise figure of 5.3 dB. In-phase and quadrature signals in the baseband are generated by mixing the received signals with the reference and filtering them with low pass filters having a cut-off frequency of 80 MHz.

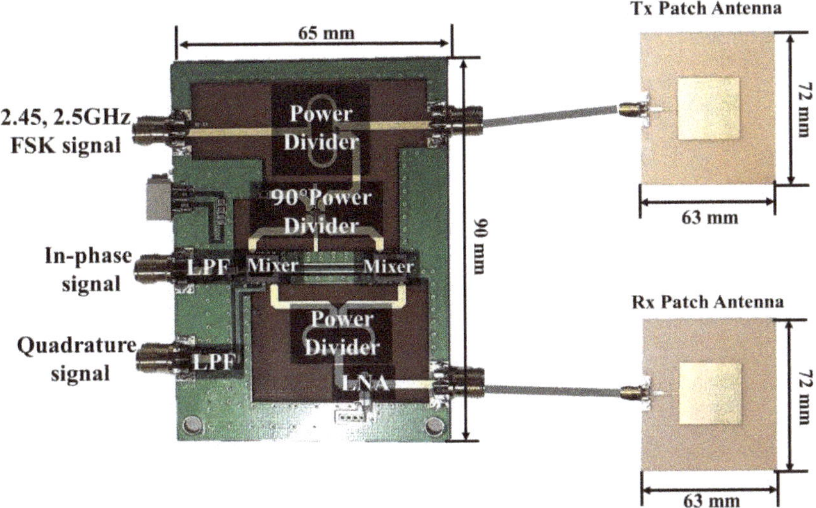

Figure 7. Implemented radar module with two patch antennas on an FR4 printed circuit board (PCB).

4. Measurement Results and Discussions

Figure 8 shows the measurement setup for obtaining both the vital signs and the distance to the subject. The switching time of the two CW frequencies was set as 0.1 s in the generator. The maximum unambiguity range was determined to be 3 m by the frequency space of 50 MHz between the two operating frequencies. By using two low-noise preamplifiers manufactured by Standford Research Systems Inc., the in-phase and quadrature signals from the module were amplified with a voltage gain of 17 dB. Signal conditioning and processing in the digital domain were implemented using NI LabVIEW and MATLAB on a personal computer after quadrature signals were simultaneously obtained with a sampling rate of 1 k samples per second from the DAQ board of NI USB-6009. A range finder (Bosch, Gerlingen, Germany) and a three-electrodes ECG sensor (Vernier Software and Technology, Beaverton, OR, USA.), were used as the reference sensors to measure the accuracies of the distance and vital signs, respectively, of the proposed FSK radar. In the surroundings, there were only fixed clutters; the moving objects (except for the subject) were limited in number. The distance to the subject was measured from 1 to 2.4 m at the intervals of 15 cm, and the vital signs were measured as respiration and heartbeat per minute for 90 s at each distance. The detection range for the demonstration is determined by considering both the nearfield effect on the radar and the maximum unambiguity range of the FSK radar. The subjects were three males in their twenties who did not suffer from cardiac diseases. They did not consume caffeine, alcohol, or nicotine, which may affect the vital signs, before measurement of the vital signs. Detailed information of the subjects is presented in Table 1.

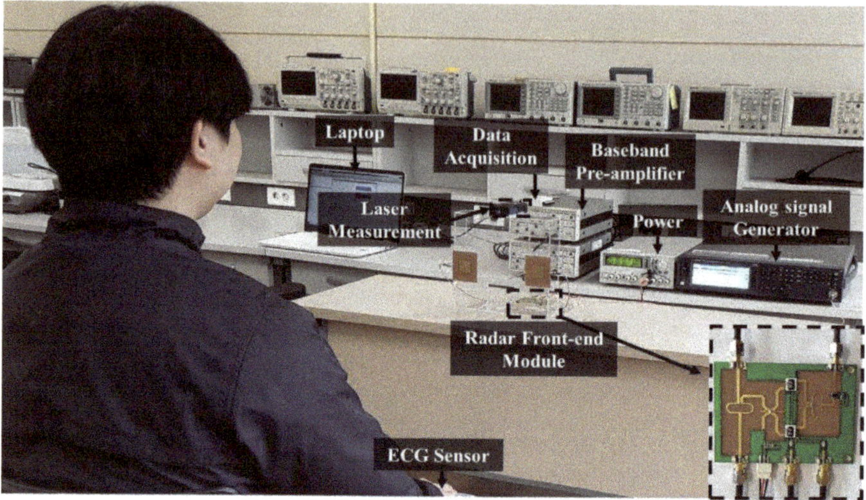

Figure 8. Measurement setup for detecting the distance and vital signs using the proposed FSK radar.

Table 1. Information relevant to the subjects.

Subjects	Gender	Age	Weight	Height	BMI
A	Male	25	90	177	28.73
B	Male	28	76	182	22.94
C	Male	25	83	172	28.06

The single-frequency in-phase signals obtained after the discrimination using the proposed envelope detection method and the signals digitally filtered by the passband from 0.1 to 0.8 Hz are shown in Figure 9. According to the operation of the CW Doppler radar, the phase information for the movement generated by the vital signs is clearly reflected by the raw data. As shown in Figure 9, the respiration is dominant in the raw data because the data are similar to the signal filtered by only the frequency band of the respiration.

The absolute distances measured by using the proposed FSK radar are shown in Figure 10. The initial phase differences in the radar front-end are calibrated with measurement results for the reference distance of 1 m, and the relative distance obtained by the radar is modified to an absolute distance after the calibration. The accuracy of the distance measurement is expressed by the root-mean-square error (*RMSE*) relative to the distance measured by the reference sensor. The *RMSE* in the distance measurement can be expressed as

$$RMSE = \sqrt{\frac{1}{n}\sum_{i=1}^{n}(d_i - rf_i)^2} \tag{10}$$

where n is the number of subjects, d_i is the absolute distance measured using the proposed radar, and rf_i is the reference distance measured using the laser-based range finder. The *RMSE* is ≤0.1 m at a distance of ≤ 1.7 m, corresponding to ≤ 6.6% of the measurement distance. The accuracy is increased with an

RMSE of 0.3 m at a distance of 1.8–2.4 m, which corresponds to 14.3% or less of the measurement distance. The increase in the measured distance error at the distances of >1.6 m is explained as follows: The phase difference of the FSK radar is determined by the vector sum of the overall signal received through the antenna. When the measurement distance increases, more information resulting from clutters is included in the received signal by the beamwidth of the receiving antenna, and it has an effect on the increase of the measurement error in the phase detection. Additionally, the SNR of the radar decreases as the measurement distance increases. The FSK radar measures the distance using the phase difference obtained from the vector sum of the overall reflected signals, including those from the surface and inside of the human body, while the reference sensor measures the absolute distance to the skin surface of the subject. The uncertainty of the phase difference obtained by the vector sum in the FSK radar results in an intrinsic error in the distance measurement for the human body. Additionally, the switching time of 0.1 s in the FSK radar is sufficiently short to neglect the changes in the vital signs, which occur less than 2 times per second. Considering the uncertainty of the phase difference and the short switching time, the error in the distance measurement of the radar is attributed to the detection accuracy of the phase in the radar. If the phase error of the FSK radar, including the uncertainty, is 5° in the measurement, the *RMSE* is 0.04 m for the frequency spacing. The distance measurement shows that vital signs obtained for each frequency at different times have a correlation at a similar level of the phase error in the measured distance.

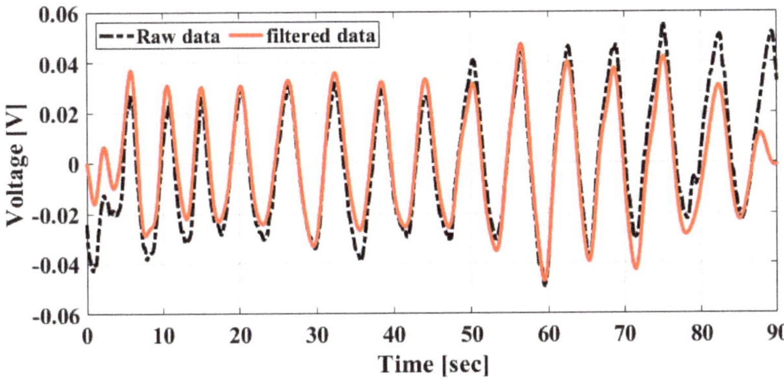

Figure 9. Raw and filtered waveforms measured at a distance of 1 m using the proposed radar.

Vital-signs detection using the proposed radar are performed by varying the subjects' position by 1 to 2.4 m at intervals of 0.15 m for a measurement time of 90 s. The respiration rate and heartbeat measured by the proposed radar are presented in the spectrum. The respiration rate per minute shown in Figure 11 for one subject is clearly measured at 12, but the heartbeat per minute (BPM) is not easily obtained in Figure 11 because of the low SNR of the heartbeat signal and the low resolution in the dynamic range to represent the respiration signal. The heartbeat signal may be measured by using high-pass filtering to reduce the respiration signal in the baseband pre-amplifier block, but this method has a limitation in increasing the measurement accuracy because it does not improve the heartbeat SNR itself. Figure 12 shows the frequency spectra of the heartbeat signals measured at a distance of 1 m using the proposed FSK radar. The raw data of the measured heartbeat show that signals due to the respiration and its harmonics might have been generated mostly around the frequency band of the heartbeat signal in the single CW radar operation. It is difficult to identify the frequency peak representing the heartbeat, because there are several peaks in the spectrum obtained at each operating frequency. The noise fluctuation level obtained at each frequency is approximately a quarter of the maximum frequency. However, the cross-correlated signals obtained using the proposed FSK radar exhibit the same frequency peak as those obtained using the reference ECG sensor and both higher SNR and lower noise level compared to those obtained at each frequency. The number of data points in

Figure 10 (based on a comparison between the single frequency and cross-correlated signals) indicates that the frequency resolution can be increased by the cross-correlation.

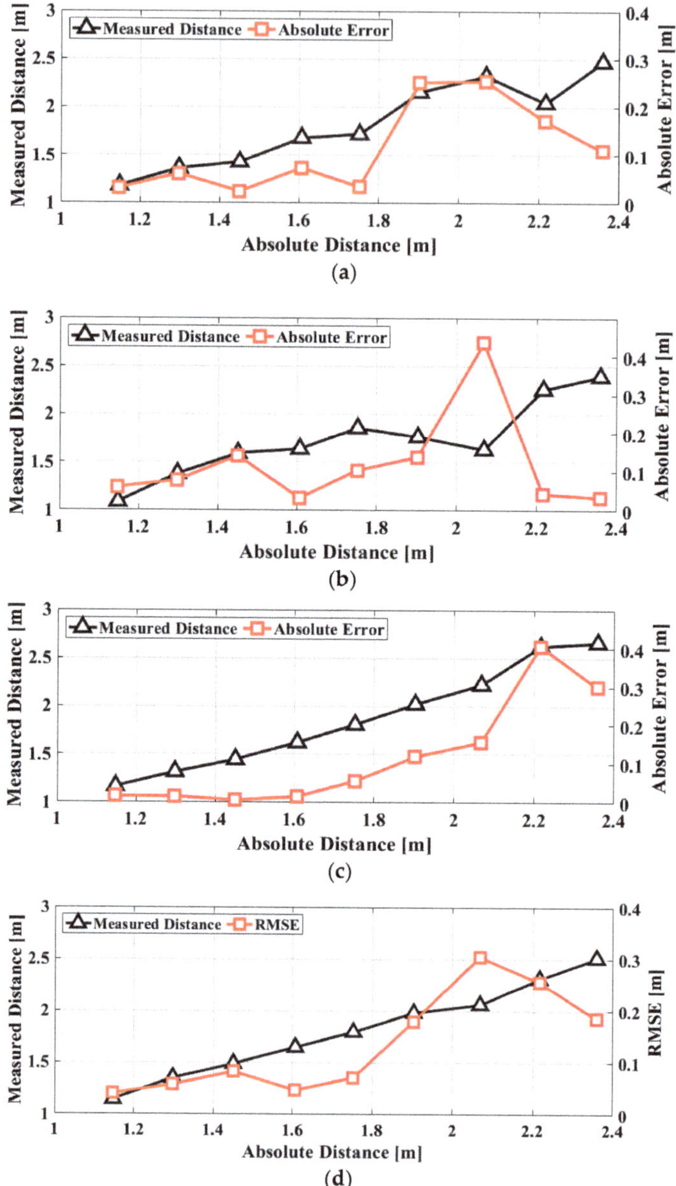

Figure 10. Measured distance and the root-mean-square error (*RMSE*) in the measurement using the proposed FSK radar with a frequency spacing of 50 MHz and switching time of 0.1 s: (**a**) Subject A; (**b**) Subject B; (**c**) Subject C; (**d**) averaged data.

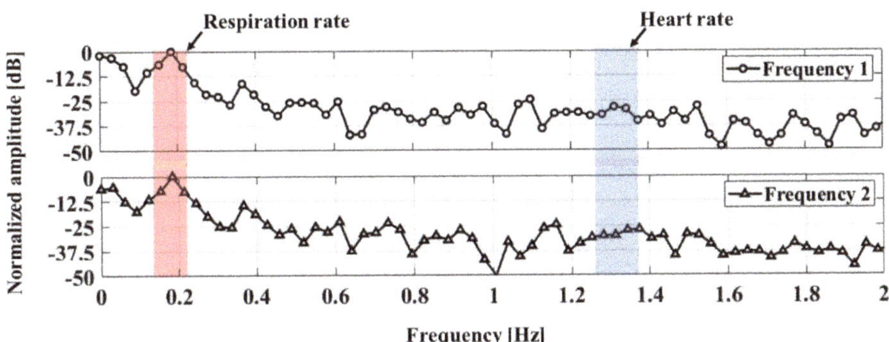

Figure 11. Spectra of the vital signals measured at a distance of 1 m using the proposed FSK radar.

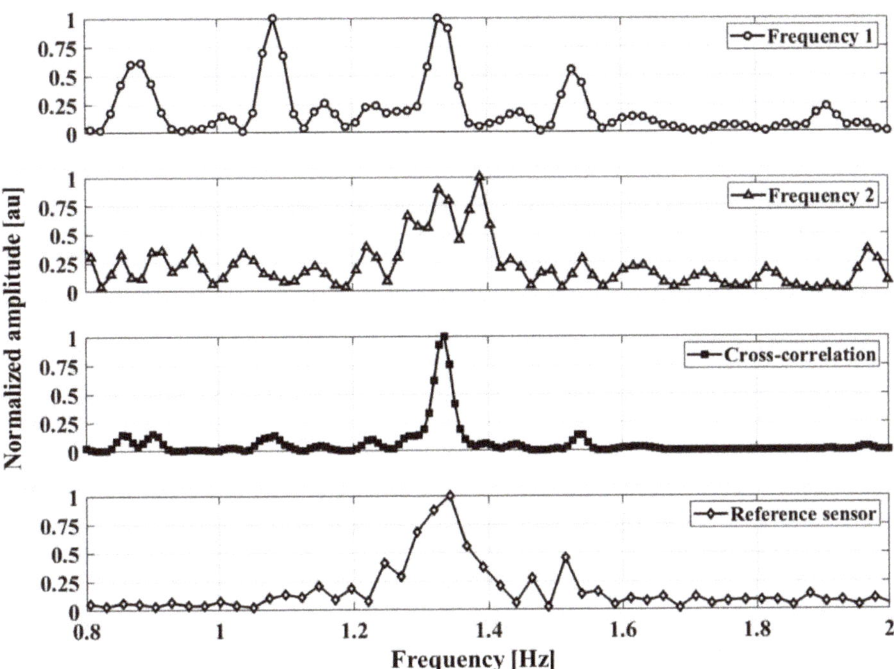

Figure 12. Spectra of the heartbeat signals measured by the proposed FSK radar at each operating frequency and with cross-correlation. The frequency peak of the cross-correlated signals is almost identical to that of the reference electrocardiogram (ECG) signals.

The measurement accuracy for a subject wearing a contact-type reference ECG sensor is evaluated using the *RMSE* and the SD. The *RMSE* in the heartbeat measurement can be expressed as

$$RMSE = \sqrt{\frac{1}{m} \sum_{i=1}^{m} (h_i - ECG_i)^2} \tag{11}$$

where m is the number of windows, h_i is the heartbeat (BPM) measured using the proposed radar, and ECG_i is the reference BPM measured using the commercialized ECG sensor. The BPM is measured for three subjects using the proposed FSK radar. The heartbeat measurement results in Figure 13 are the *RMSEs* and SDs of the subjects depending on the measurement distance. The results of the proposed

FSK radar typically exhibit a smaller *RMSE* and SD than those measured at individual operating frequencies, i.e., those for the CW Doppler radar. When the measurement distance is increased, both the *RMSE* and SD of the heartbeat measurement increase corresponding to the reduction in the SNR. At a distance of 2.05 m, the accuracy obtained from the cross-correlated signals is similar to that obtained from the single-frequency signals. This indicates that the accuracy improvement due to the cross-correlation for the proposed FSK radar might not be significant when the effect of the low SNR and the uncertainty of the phase measurement increase owing to the increase in distance from the radar. Table 2 presents the average *RMSEs* and SDs for all BPM measurements for each subject. As shown, the cross-correlated signals had small *RMSEs* and SDs for all the subjects. In the proposed FSK radar, the average *RMSEs* and SDs of the heartbeat obtained using the cross-correlation method are improved by 2.42 and 2.36 BPM, respectively, compared to those measured at each frequency. With regard to the operating principle and the procedure, heartbeat measurement with only a single frequency using the proposed radar is identical to that using the CW Doppler radar. The measurement results demonstrate that the proposed FSK radar with the cross-correlation in a single hardware configuration is advantageous for improving the accuracy of vital-signs detection compared to the CW Doppler radar. Table 3 summarizes the performance comparison of the proposed radar with the radars from our previous studies, which were employed for vital-signs detection in similar measurement surroundings.

Table 2. Average root-mean-square error (*RMSE*) and standard deviation (SD) of overall heartbeat per minute measured for each subject by using the proposed FSK radar.

Subject	*RMSE* in BPM			SD in BPM		
	2.45 GHz	2.5 GHz	CC [1]	2.45 GHz	2.5 GHz	CC [1]
A	11.970	9.347	6.260	11.814	9.324	6.264
B	7.160	7.478	5.599	7.032	7.524	5.640
C	5.037	5.609	4.393	5.034	5.550	4.422
Average	8.562	7.632	5.472	8.550	7.602	5.460

[1] The cross-correlation method.

Table 3. Performance comparison of the proposed radar with the radars from our previous studies, which were used for vital-signs detection in a similar measurement environment.

	[5]	[10]	[14]	This Work
Type	CW	CW	CW	FSK
Operating Frequency (GHz)	2.45	0.915	2.45	2.45 & 2.5
Distance to target (m)	0.4	0.2–1.4	1.0	1.0–2.4
Signal Processing	Peak detection	Fast-Fourier transform	Wavelet transform	Cross-correlation
Detectable Information	Respiration, heartbeat	Respiration, heartbeat	Respiration, heartbeat	Respiration, heartbeat, distance
Mean absolute error in HR [1] (%)	3.22	1.37	3.93	6.00
Reference ECG sensor		Three-electrodes sensor manufactured by Vernier Software and Technology		

[1] Calculated by using the average error of heartrate (BPM) measured in all detectable ranges.

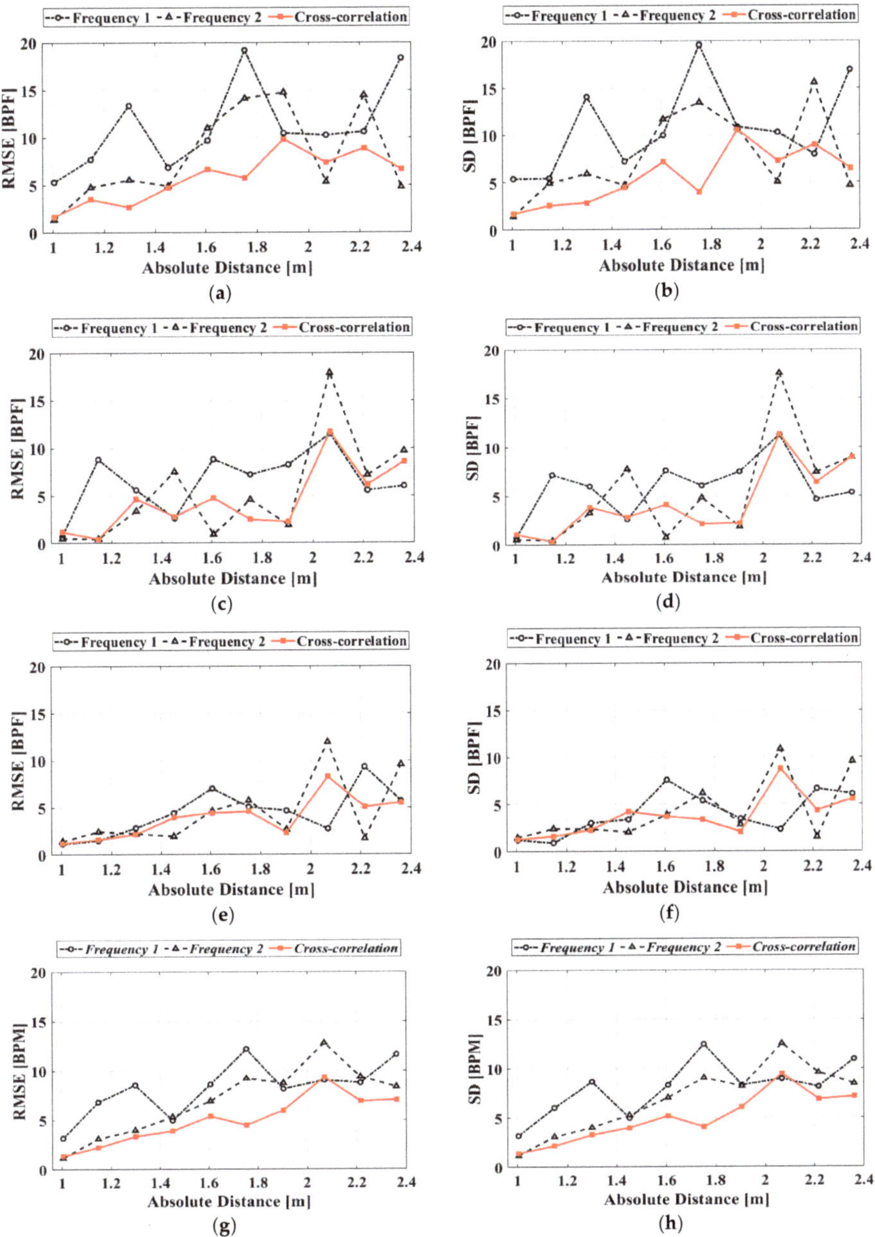

Figure 13. Measured accuracies of the signals obtained at each operating frequency and signals processed with the cross-correlation method in the proposed FSK radar depending on the distance to the subjects: (**a**) Root-mean-square error (*RMSE*) of subject A; (**b**) standard deviation (SD) of subject A; (**c**) *RMSE* of subject B; (**d**) SD of subject B; (**e**) *RMSE* of subject C; (**f**) SD of subject C; (**g**) *RMSE* of the averaged data; (**h**) SD of the averaged data.

5. Conclusions

A remote vital signs detector based on FSK radar with the cross-correlation method implemented in a single radar architecture is presented for accurately detecting heartbeat signals. The proposed FSK radar can increase the SNR of vital signs and the frequency resolution in the FFT spectrum by using cross-correlation between the phase differences individually obtained at two operating frequencies. An envelope detection method for controlling the frequency-switching signal for FSK operation is proposed for phase-difference discrimination, which is mandatory for obtaining vital signs at each operating frequency. Distance measurements for human subjects shows that the vital signs obtained at each operating frequency of the proposed FSK radar can be correlated with each other while producing an acceptable distance error. The measurement results for the number of heartbeats per minute for subjects at a distance of 1–2.4 m demonstrate that the proposed radar can improve the vital-signs detection accuracy and SNR via cross-correlation.

Author Contributions: Conceptualization, J.-R.Y.; methodology, J.Y.S., J.-H.P., and J.-R.Y.; software, J.Y.S. and J.-H.P.; validation, J.Y.S. and J.-R.Y.; formal analysis, J.Y.S., J.-H.P., and J.-R.Y.; investigation, J.Y.S., J.-H.P., and J.-R.Y.; resources, J.-R.Y.; data curation, J.Y.S.; writing—original draft preparation, J.Y.S. and J.-R.Y.; writing—review and editing, J.-R.Y.; visualization, J.Y.S.; supervision, J.-R.Y.; project administration, J.-R.Y.; funding acquisition, J.-R.Y. All authors have read and agreed to the published version of the manuscript.

Funding: This work was partly supported by the Institute of Information & Communications Technology Planning & Evaluation (IITP) grant funded by the Korea government (MSIT) (No. 2019-0-00138) and the Basic Science Research Program through the National Research Foundation of Korea (NRF) funded by the Korea government (MSIT) (No. 2017R1C1B2002285).

Acknowledgments: The authors thank Hyun-Jun Ji, who assisted with the vital-signs detection, and Ju-Hee Son, who designed the patch antennas.

Conflicts of Interest: The authors declare no conflict of interest.

References

1. Lin, J.C. Noninvasive microwave measurement of respiration. *Proc. IEEE* **1975**, *63*, 1530. [CrossRef]
2. Chen, K.-M.; Misra, D.; Wang, H.; Chuang, H.-R.; Postow, E. An X-band microwave life-detection system. *IEEE Trans. Biomed. Eng.* **1986**, *33*, 697–701.
3. Chen, K.-M.; Huang, Y.; Zhang, J.; Norman, A. Microwave life-detection systems for searching human subjects under earthquake rubble or behind barrier. *IEEE Trans. Biomed. Eng.* **2000**, *47*, 105–114. [CrossRef] [PubMed]
4. Lin, F.; Zhuang, Y.; Song, C.; Wang, A.; Li, Y.; Gu, C.; Li, C.; Xu, W. SleepSense: A noncontact and cost-effective sleep monitoring system. *IEEE Trans. Biomed. Circuits Syst.* **2016**, *11*, 189–202. [CrossRef]
5. Kim, J.-Y.; Park, J.-H.; Jang, S.-Y.; Yang, J.-R. Peak detection algorithm for vital sign detection using Doppler radar sensors. *Sensors* **2019**, *19*, 1575. [CrossRef]
6. Brüster, C.; Antink, C.H.; Wartzek, T. Ambient and unobtrusive cardiorespiratory monitoring techniques. *IEEE Rev. Biomed. Eng.* **2015**, *8*, 30–43. [CrossRef]
7. Shang, N.; Xin, K.; Yu, C.; Wang, L. Design of direct wave cancellation system for high-frequency CW radar. *J. Eng.* **2019**, *2019*, 6660–6663. [CrossRef]
8. De Groote, A.; Wantier, M.; Chéron, G.; Estenne, M.; Paiva, M. Chest wall motion during tidal breathing. *J. Appl. Physiol.* **1997**, *83*, 1531–1537. [CrossRef]
9. Ramachandran, G.; Singh, M. Three-dimensional reconstruction of cardiac displacement patterns on the chest wall during the P, QRS and T-segments of the ECG by laser speckle interferometry. *Med. Biol. Eng. Comput.* **1989**, *27*, 525–530. [CrossRef]
10. Park, J.-H.; Jeong, Y.-J.; Lee, G.-E.; Oh, J.-T.; Yang, J.-R. 915-MHz continuous-wave Doppler radar sensor for detection of vital signs. *Electronics* **2019**, *8*, 561. [CrossRef]
11. Guanghao, S.; Matsui, T. Rapid and stable measurement of respiratory rate from Doppler radar signals using time domain autocorrelation model. In Proceedings of the 2015 37th Annual Int. Conf. of the IEEE Eng. in Medicine and Biology Society, Milan, Italy, 25–29 August 2015.
12. Li, M.; Lin, J. Wavelet-transform-based data-length-variation technique for fast heart rate detection using 5.8-GHz CW Doppler radar. *IEEE Trans. Microw. Theory Techn.* **2018**, *66*, 568–576. [CrossRef]
13. Iyer, B.; Garg, M.; Pathak, N.P.; Ghosh, D. Concurrent dual-band RF system for human respiration rate and heartbeat detection. In Proceedings of the 2013 IEEE Conf. on Information & Communication Technologies, Thuckalay, Tamil Nadu, India, 11–12 April 2013.

14. Choi, C.-H.; Park, J.-H.; Lee, H.-N.; Yang, J.-R. Heartbeat detection using a Doppler radar sensor based on the scaling function of Wavelet transform. *Microw. Opt. Techn. Lett.* **2019**, *61*, 1792–1796. [CrossRef]

15. Yang, J.-R.; Hong, S. A 24-GHz radar sensor with a six-port network for short-range detection. *Microw. Opt. Techn. Lett.* **2014**, *56*, 2634–2637. [CrossRef]

16. Park, J.-H.; Yang, J.-R. Two-tone CW Doppler radar based on envelope detection method. *Microw. Opt. Techn. Lett.* **2020**, *62*, 3146–3150. [CrossRef]

17. Hu, W.; Zhao, Z.; Wang, Y.; Zhang, H.; Lin, F. Noncontact accurate measurement of cardiopulmonary activity using a compact quadrature Doppler radar sensor. *IEEE Trans. Biomed. Eng.* **2014**, *61*, 725–735. [CrossRef] [PubMed]

18. Droitcour, A.D.; Boric-Lubecke, O.; Lubecke, V.M.; Lin, J.; Kovacs, G.T. Range correlation and I/Q performance benefits in single-chip silicon Doppler radars for noncontact cardiopulmonary monitoring. *IEEE Trans. Microw. Theory Techn.* **2004**, *52*, 838–848. [CrossRef]

19. Siddiq, K.; Hobden, M.K.; Pennock, S.R.; Watson, R.J. Phase noise in FMCW radar systems. *IEEE Trans. Aerosp. Electron. Syst.* **2019**, *55*, 70–81. [CrossRef]

20. Smith, S. *Digital Signal Processing: A Practical Guide for Engineers and Scientists*; Newnes: Oxford, UK, 2003; pp. 136–140.

21. Wang, J.; Wang, X.; Chen, L.; Huangfu, J.; Li, C.; Ran, L. Noncontact distance and amplitude-independent vibration measurement based on an extended DACM algorithm. *IEEE Trans. Instrum. Meas.* **2013**, *63*, 145–153. [CrossRef]

22. Yang, J.-R.; Kim, D.-W.; Hong, S. A calibration method of a range finder with a six-port network. *IEEE Microw. Wireless Compon. Lett.* **2007**, *17*, 549–551. [CrossRef]

23. Singh, A.; Gao, X.; Yavari, E.; Zakrzewski, M.; Cao, X.; Lubecke, V.; Boric-Lubecke, O. Data-based quadrature imbalance compensation for a CW Doppler radar system. *IEEE Trans. Microw. Theory Techn.* **2013**, *61*, 1718–1724. [CrossRef]

24. Li, C.; Zhao, H.; Xi, F. Accurate DC offset calibration of Doppler radar via non-convex optimization. *Electron. Lett.* **2015**, *51*, 1282–1284.

25. Pratt, V. Direct least-squares fitting of algebraic surfaces. *ACM SIGGRAPH Comput. Graph.* **1987**, *21*, 145–152. [CrossRef]

26. Fan, T.; Ma, C.; Gu, Z.; Lv, Q.; Chen, J.; Ye, D.; Huangfu, J.; Sun, Y.; Li, C.; Ran, L. Wireless hand gesture recognition based on continuous-wave Doppler radar sensors. *IEEE Trans. Microw. Theory Techn.* **2016**, *64*, 4012–4020. [CrossRef]

27. Li, C.; Lin, J. Complex signal demodulation and random body movement cancellation techniques for non-contact vital sign detection. In Proceedings of the 2008 IEEE Int. Microw. Symp., Atlanta, GA, USA, 15–20 June 2008; pp. 567–570.

28. Wang, J.; Karp, T.; Muñoz-Ferreras, J.M.; Gómez-García, R.; Li, C. A spectrum-efficient FSK radar technology for range tracking of both moving and stationary human subjects. *IEEE Trans. Microw. Theory Techn.* **2019**, *67*, 5406–5416. [CrossRef]

 *sensors*

Article

The Optimal Selection of Mother Wavelet Function and Decomposition Level for Denoising of DCG Signal

Young In Jang , Jae Young Sim, Jong-Ryul Yang * and Nam Kyu Kwon *

Department of Electronic Engineering, Yeungnam University, Gyeongsan, Gyeongbuk 38541, Korea;
owlisie@yu.ac.kr (Y.I.J.); ja2922@yu.ac.kr (J.Y.S.)
* Correspondence: jryang@yu.ac.kr (J.-R.Y.); namkyu@yu.ac.kr (N.K.K.); Tel.: +82-53-810-2495 (J.-R.Y.);
+82-53-3095 (N.K.K.)

Abstract: The aim of this paper is to find the optimal mother wavelet function and wavelet decomposition level when denoising the Doppler cardiogram (DCG), the heart signal obtained by the Doppler radar sensor system. To select the best suited mother wavelet function and wavelet decomposition level, this paper presents the quantitative analysis results. Both the optimal mother wavelet and decomposition level are selected by evaluating signal-to-noise-ratio (SNR) efficiency of the denoised signals obtained by using the wavelet thresholding method. A total of 115 potential functions from six wavelet families were examined for the selection of the optimal mother wavelet function and 10 levels (1 to 10) were evaluated for the choice of the best decomposition level. According to the experimental results, the most efficient selections of the mother wavelet function are "db9" and "sym9" from Daubechies and Symlets families, and the most suitable decomposition level for the used signal is seven. As the evaluation criterion in this study rates the efficiency of the denoising process, it was found that a mother wavelet function longer than 22 is excessive. The experiment also revealed that the decomposition level can be predictable based on the frequency features of the DCG signal. The proposed selection of the mother wavelet function and the decomposition level could reduce noise effectively so as to improve the quality of the DCG signal in information field.

Keywords: doppler cardiogram; wavelet transform; denoising; mother wavelet function; decomposition level; signal decomposition; signal-to-noise-ratio

Citation: Jang, Y.I.; Sim, J.Y.; Yang, J.-R.; Kwon, N.K. The Optimal Selection of Mother Wavelet Function and Decomposition Level for Denoising of DCG Signal. *Sensors* **2021**, *21*, 1851. https://doi.org/10.3390/s21051851

Academic Editor: Manuel Blanco-Velasco

Received: 28 January 2021
Accepted: 3 March 2021
Published: 6 March 2021

Publisher's Note: MDPI stays neutral with regard to jurisdictional claims in published maps and institutional affiliations.

1. Introduction

Cardiovascular diseases (CVDs) are the leading and rapidly increasing cause of death in modern society. They caused 32.1% of annual global deaths in 2015 and the occurrence is growing [1]. According to this trend, research on heartbeat detection and heart rate variability analysis has been gaining increasing attention. As the basic method of the heart signal recording, the electrocardiogram (ECG) test is used for measuring the electrophysiological signal of the heart activity to diagnose symptoms of cardiovascular problems [2]. The ECG test can improve the diagnosis and therapy of the most prevalent cardiac diseases if it does not interfere with daily activities [3]. However, some requirements of the ECG test make subjects uncomfortable. For example, subjects are required to attach multiple electrodes on their skin and remain in static state for quite some time [4]. In addition as the abnormal heartbeat signal appears for a very short time, the ECG test could miss these short and temporary symptoms because this test only records in limited circumstances.

Doppler radar sensors detect the electrophysiological signal of the heart and the variation of blood vessel movement by differences of the electromagnetic signal that is a few dozens of centimeters away from a subject [5–8]. Due to the non-contactable and flexible measurement condition, as opposed to ECG sensing, the Doppler radar sensors are studied for the alternative of ECG electrodes [9]. However, the noise occurs in the output signal obtained by the radar sensors due to the limitation of the non-contact measurement. Since the noise has a negative effect on data acquisition and processing, it leads to a decrease in information accuracy. Therefore, the method for signal denoising is essential and a considerable amount of research related to denoising has been studied [10–14]. Among them, the noise reduction methods using the wavelet analysis are widely studied not only for the radar signal but also for various multi-dimensional signals [15–20].

Noise reduction methods using the wavelet transform and thresholding can improve the quality of the Doppler cardiogram (DCG) and overcome the limits of remote sensor. To successfully enhance signal quality, the noise reduction technique should satisfy the performance of removing the noise components and maintaining useful signal components. In comparison to other noise removal methods such as average filtering or frequency smoothing using the fast Fourier transform, the wavelet denoising method shows a superior performance for preserving the features of the original signal when removing noise components. In particular, it is effective in regards to the problem of non-stationary signals such as a heart signal, respiration signal, and ocular artifacts [21–25]. The multiresolution decomposition of the wavelet transform makes it possible to maintain the original information during the process of removing noise components. Meanwhile, since various scaling factors and features of the wavelet functions are available, the optimal selection procedure of the mother wavelet function is a significant step to the wavelet denoising method [26,27]. Traditionally, the mother wavelet function is selected to represent the characteristics of the signal by empirical research [28,29]. Since this method does not provide the actual optimal selection of the mother wavelet [30], studies of the mother wavelet selection for various signals such as the electroencephalography (EEG) signal, vibration signal of turbine, human voice signal, and many kinds of radar signal have been occurring [26,28,31–36]. However, to the best of the author's knowledge, studies on the optimal selection of the mother wavelet function and the decomposition level for the denoising of the DCG signal are rarely reported. Though selection of the mother wavelet function and the decomposition level on other radar signals were presented [35,36], distinctive features of each kind of radar signal require an individualized selection respectively, which is the motivation behind this study.

In this study, the optimal selection of the mother wavelet function and the decomposition level for the cardiac signal obtained by the Doppler radar sensors is proposed with quantitative analysis results. The result of the denoising process is defined with the signal-to-noise-ratio (SNR) efficiency of the denoised signal. The candidates for the optimal mother wavelet function are composed of 115 functions from six different wavelet families and the decomposition level has 10 candidates from 1 to 10. To determine the optimal mother wavelet function and the optimal decomposition level, the denoising process is repeated to examine all candidates and then all results are accumulated. The optimal choice of the mother wavelet function and the decomposition level is selected among the accumulated results which perform the highest SNR efficiency. The experiment is executed in two steps to find out the optimal selection of the mother wavelet function and the decomposition level. For the first step, the optimal mother wavelet function is determined by using the arbitrary wavelet decomposition level. Then the optimal decomposition level is obtained by using the optimal mother wavelet function which is selected at the first step. In this study, the arbitrary decomposition level in the first step is predicted based on the distribution of the signal components in the frequency domain. With these experiments, it is found that the length of the mother wavelet function is one of the key determinants for the optimal selection. The noisy signals were generated by adding white Gaussian noise to the original signals. The DCG dataset has 28 recordings that have been recorded for 160 with a sampling frequency of 1000 Hz.

From the experimental results, it can be seen that the optimal mother wavelet function is found as db9 and sym9, and the optimal wavelet decomposition level is seven respectively. The main contributions of this study can be described as follows:

- The optimal selection of the mother wavelet function and the decomposition level for the DCG (Doppler cardiogram) is introduced in this study;
- The optimal wavelet decomposition level is predicted based on the distribution of the signal components in frequency domain;

- The new criterion is suggested to evaluate both the denoising performance and the denoising efficiency in once.

2. Materials and Methods

2.1. Wavelet Transform

The wavelet transform, a linear transformation with the mother wavelet function, is an efficient signal analysis method for both time and frequency resolution [22,23,37]. The wavelet transform has advantages over traditional Fourier transforms for accurately deconstructing and reconstructing finite, non-periodic, and non-stationary signals such as DCG signals. The wavelet transform decomposes the input signal into detail coefficients (cD) and approximation coefficients (cA) that can be defined as high frequency coefficients ($y_{high}[n]$) and low frequency coefficients ($y_{low}[n]$), respectively [38]. The frequency band coefficients of the wavelet transform are expressed mathematically as follows [38]:

$$y_{high}[n] = \sum_{i=-\infty}^{\infty} s[i]h[2n-i], \tag{1}$$

$$y_{low}[n] = \sum_{i=-\infty}^{\infty} s[i]g[2n-i], \tag{2}$$

where i is a sampling data point; n is the number of the sampling data; $s[i]$ is the discrete radar signal with noise; and $g[2n-i]$ and $h[2n-i]$ are low-pass and high-pass filters that vary depending on the mother wavelet function [39]. With frequency filters, the wavelet transform enables one to extract the particular frequency band from the original signal [40,41]. Figure 1 shows the x level decomposition process into the approximation coefficients cA and the detail coefficients cD of the N Hz signal:

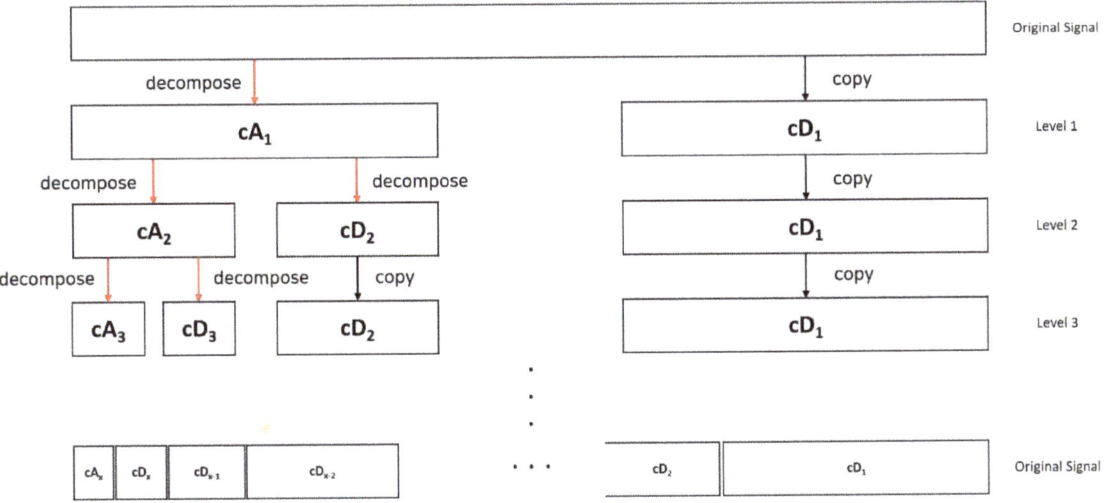

Figure 1. Illustration of the x level decomposition process.

The wavelet function is composed with the scaled and translated copies of the scaling function $\phi(x)$ and the mother wavelet function $\psi(x)$ [26,38] which is a continuously differentiable function with compact support. Since there are many different types of mother wavelet functions, finding the optimal mother wavelet is essential for the best performance not only in denoising but also in other signal processing. The coefficients of the discrete wavelet transform represent the projection of the signal over a set of basis functions generated as the translation and dilation of the mother wavelet function and the scaling function [42]. More specifically, the low-pass coefficients are related to the scaling function and the high-pass coefficients are related to the mother wavelet function as in Equations (3)–(5) [38]:

$$g(h) = (-1)^n h(1 - n),$$

$$\phi(x) = \sum_n h(n) \sqrt{2} \phi(2x - n),$$

$$\psi(x) = \sum_n g(n) \sqrt{2} \phi(2x - n).$$

The low-pass and high-pass filters are distinctive variables for each mother wavelet function. Since the selection of the mother wavelet function and the decomposition level affects the noise reduction performance, the suitable mother wavelet function and the decomposition level could be optimizing the performance of the processing methods based on the wavelet transform.

2.2. Doppler Cardiogram Signal Dataset and Recording Procedure

A total of 28 DCG signals from 11 subjects were examined in this study. The datasets were recorded from healthy control subjects composed of 6 females and 5 males aged 24.08 $\pm$ 2.35 years (mean $\pm$ standard deviation). The subjects were recruited from the Electronic Engineering department of Yeungnam University. The subjects have no history of CVDs. Each signal data was recorded for 160 s and 1 k samples per second. The measurement system was constructed with a Doppler radar sensor, which measured the cardiogram of the subject at a distance of 30 cm. Table 1 shows the socio-demographic and clinical data of the control subjects. Permission to use the DCG data was granted from all subjects.

Table 1. Sociodemographic data of the control subjects (Age in years, mean $\pm$ standard deviation, SD.

Demographic and Clinical Features	Control
Number	11
Age	24.08 $\pm$ 2.35
Female/Male	6 F/5 M

The DCG signals were detected by using the radar sensor module with two discrete patch antennas, which was implemented on an FR4 printed circuit board with a thickness of 0.6 mm, as shown in Figure 2 [43]. A total of 5.8 GHz of signals were generated by a voltage-controlled oscillator (HMC431LP4, Analog Devices) with an output power of 7.8 dBm radiating from the transmitting antenna with a directivity of 4.0 dBi. The received signals modulated by respiration and heartbeat were incident by using the receiving antenna, and they are down-converted to baseband signals by using quadrature mixers with low-pass filters with a cut-off frequency of 500 MHz. The overall noise figure in the receiver path was calculated to approximately 1.9 dB. The baseband signals were collected on a PC by using the data acquisition board with a sampling frequency of 1 kHz. Vital signs were obtained by filtering the baseband signals with a high-order band-pass filter in the digital domain. The multi-phase mixing in the radar module was used in the DC offset compensation in the received signals to improve the accuracy of the phase demodulation in signal processing [43]

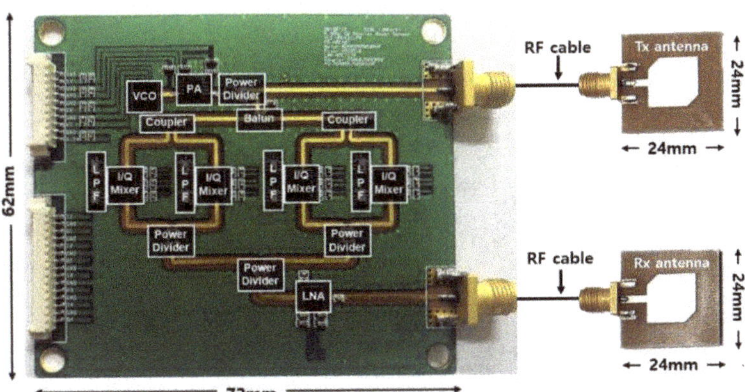

Figure 2. A 5.8 GHz continuous-wave Doppler radar sensors for the measurement of vital signs [43].

Figure 3 shows the experimental setup for vital signs detection with a 5.8 GHz CW Doppler radar. The vital signs detected by the radar, which is located at a distance of 30 cm from the thorax of the subject and fixed on the bottom of the table above the subject, were obtained by using the data acquisition board (NI USB-6366, National Instruments, Austin, TX, USA) at a sampling rate of 1 kHz. A three-electrode ECG sensor (Vernier Software and Technology, Beaverton, OR, USA.), which was used as the reference sensor for heartbeat detection of the radar, was obtained at a sampling rate of 200 Hz. Heartbeat signals simultaneously obtained by the radar and ECG sensors were compared using NI LabVIEW and MATLAB on a computer. The vital signals of 11 subjects (5 males and 6 females in their 20 s) lying on the bed were continuously measured for 160 s. The movement of the subjects were limited to minimize the effect of the noise generated by motion on the ECG and radar sensors.

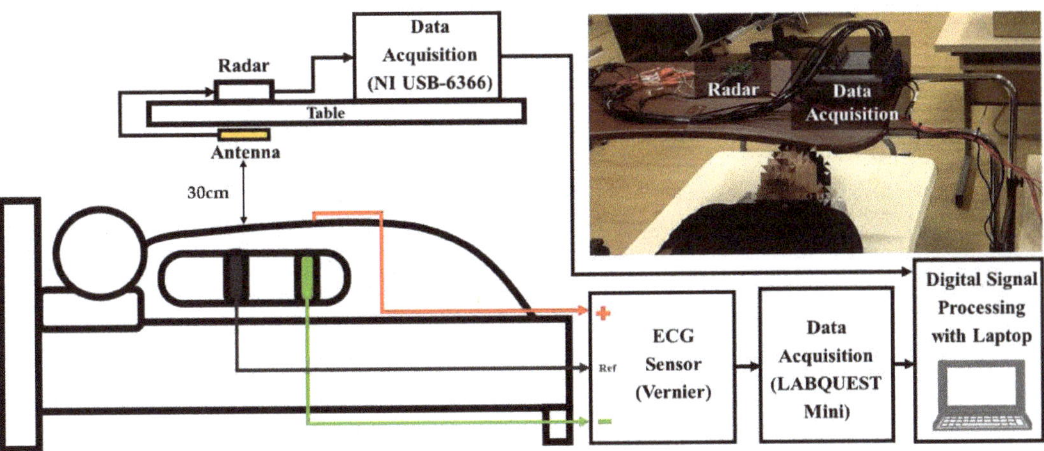

Figure 3. Block diagram of the experimental setup for vital signal detection using the 5.8 GHz CW radar.

Though the ECG signals were not used in this study, ECG and DCG signals were measured from each subject once due to the structure of the sensor system. The corresponding matches of the ECG and DCG were obtained as a dataset for the future work of cardiogram analysis.

2.3. Evalution Measures

In this section, the optimal choice will be derived based on the quantitative experimental results. Since the SNR is one of the measures that evaluates the denoising performance [44,45], the following performance indexes were used to evaluate the denoised signal with j and k.

$$SNR_{j,\,k} = 10log_{10}\frac{\sum_{i=1}^{N} s^2[i]}{\sum_{i=1}^{N}\left|s[i] - \hat{s}_{j,\,k}[i]\right|}, \tag{6}$$

$$\eta(k) = log_{10}((s_l - l(m[k])) \times l(m[k])), \tag{7}$$

$$eff(j,\,k) = \frac{SNR_{j,k}}{\eta(k)} \tag{8}$$

where $\hat{s}_{j,k}[i]$ and $SNR_{j,\,k}$ in (10) denote the denoised signal and the SNR when the decomposed level is j and the wavelet function index is k. The $\eta(k)$ (11) is the execution complexity of the wavelet transform in case the wavelet function index is k. The $\eta(k)$ is defined as the common logarithm value of the number of multiplication operations required for the wavelet transform that can be obtained by the length of the signal s_l and the length of the mother wavelet function $l(m[k])$. For the denoising process, the higher SNR value shows that a better performance and lower execution complexity represents higher efficiency [46]. Therefore, the new criterion $eff(j,\,k)$ (12) is proposed in this study to evaluate denoising performance efficiency.

3. Experiment

3.1. Additive White Gaussian Noise

Additive white Gaussian noise is the fundamental noise model in the information signal processing with three important characteristics manifested in its term. First, additive means the way in which noise joins in the signal. The noisy signal $s_N[n]$ is generated by adding the noise components $N[n]$ to the original signal $s[n]$ [47]. This process is described in Figure 4 and the mathematical expression is described as below:

$$s_N[n] = s[n] + N[n]. \tag{9}$$

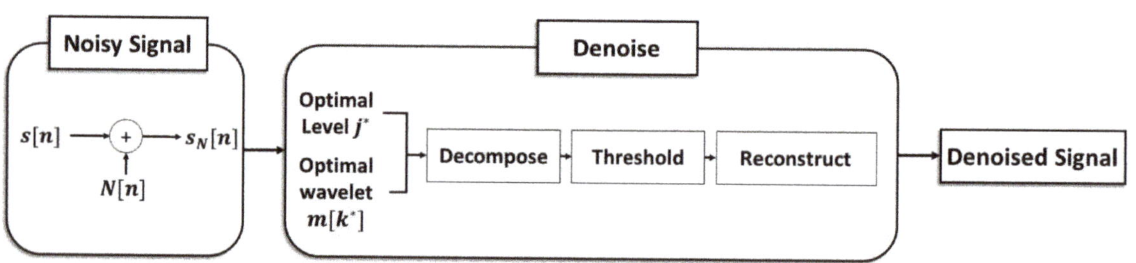

Figure 4. The block diagram of the denoising method.

Second, white denotes the power spectrum density of the noise is a constant value across the frequency range and the noise exists at almost every frequency [48]. Finally, Gaussian represents that the distribution of the noise is the Gaussian random process. The additive white Gaussian noise consists of normal distribution in the time domain and independent and identical distribution (i.i.d) in the frequency domain [49]. In this study, the additive white Gaussian noise components follow the distribution of $N(0, \sigma)$, which is copied to the electromagnetic noise distribution of the original signal. Since the distribution is identical, the generated noisy signal $s_N[n]$ can represent comparable features with the real world noisy signal [50]. In the experiment, the white Gaussian noise is added to each original DCG signal with a SNR of 10 dB by the signal processing tool that is provided from Mathworks MATLAB.

3.2. Denoising Process with Wavelet Transform and Thresholding

To improve the quality of the noisy signal, the denoising process should hold the real data factors and remove the additive noise components [27]. In the previous studies, the noise reduction methods are generally based on the model simulation or spectral analysis, in which it is difficult to keep the complicated features of the signal [27,51–53]. On the other hand, the wavelet decomposition thresholding method can control the noisy signal to maintain the original signal components and separate the noise components more precisely. In this study, the denoising method is composed of three steps as follows [54,55]:

1. Wavelet Decomposition Select the decomposition level j and the mother wavelet function $m[k]$. Produce the wavelet coefficients through discrete wavelet transform with these two factors.
2. Thresholding Set the threshold value, which is calculated by the wavelet coefficients. Threshold the decomposed wavelet coefficients.
3. Wavelet Reconstruction Reconstruct the wavelet coefficients after thresholding using j and $m[k]$.

Figure 4 describes the whole denoising process as a block diagram with an illustration of the three steps of the denoising method. j denotes the decomposition level of 1 to 10, $m[k]$ is the mother wavelet function, and k is the function index from 1 to 115. Meanwhile, at the thresholding step, the soft thresholding is used for its superior thresholding performance to the hard thresholding [56]. There are various threshold methods for noise elimination such as Empirical Bayes [57], Block James–Stein [58], False Discovery Rate [59], Minimax Estimation [60], Stein's Unbiased Risk Estimate [61], and Universal Threshold [62,63]. As the threshold method affects the denoising performance, studies on the powerful threshold determination method are kept reported [64,65]. In this study, the Universal Threshold is selected for its simple operation and powerful performance. The mathematical expressions for the soft thresholding are described in Equations (10)–(12) [51]:

$$y_s(c) = \left\{ \begin{array}{ll} sgn(c)\cdot(|c| - T_U), & |c| > T_U \\ 0, & |c| \leq T_U \end{array} \right\}, \tag{10}$$

$$T_U = \hat{\sigma}\sqrt{2log(N)}, \tag{11}$$

$$\hat{\sigma} = MAD/0.6745, \tag{12}$$

where c is the wavelet coefficient of the decomposed noisy signal [65,66] and T_U is the universal threshold value proposed by Donoho and Johnstone [51,67]. N is the number of samples in the signal and MAD is the median absolute deviation of the wavelet coefficients. The meaning of $\hat{\sigma}$ is the estimate of the standard deviation of the noise. Following these equations, the threshold value is produced with the median value of the wavelet coefficients and the number of the sampling ratio in the signal.

3.3. The Process of the Optimal Selection of the Mother Wavelet Function and the Decomposition Level

As the decomposition level and mother wavelet function have various candidates, the optimal selection not only evaluates a superior performance of denoising but also rates the execution efficiency to reduce unnecessary calculation. Table 2 and Figure 5 described the process of the optimal selection. First, the denoising process is repeated to accumulate the evaluation results for all candidates. Then, the optimal decomposition level and wavelet function index are obtained by finding the arguments of the maxima from the accumulated results. In Figure 5, j^* denotes the optimal decomposition level and $m[k^*]$ denotes the optimal mother wavelet function, where k^* is the function index of the optimal mother wavelet.

Table 2. The table of algorithm step for the analysis process of the optimal selection.

Algorithm Step
• Accumulate the denoising result using 115 mother wavelet functions and decomposition level 7 (prediction).
• Find out the most efficient mother wavelet function $m[k^*]$.
• Accumulate the denoising result using the optimal mother wavelet function $m[k^*]$ and 10 decomposition levels.
• Find out the most efficient wavelet decomposition level j^*.
• Set the optimal selection of the mother wavelet function $m[k^*]$ and the wavelet decomposition level j^*.

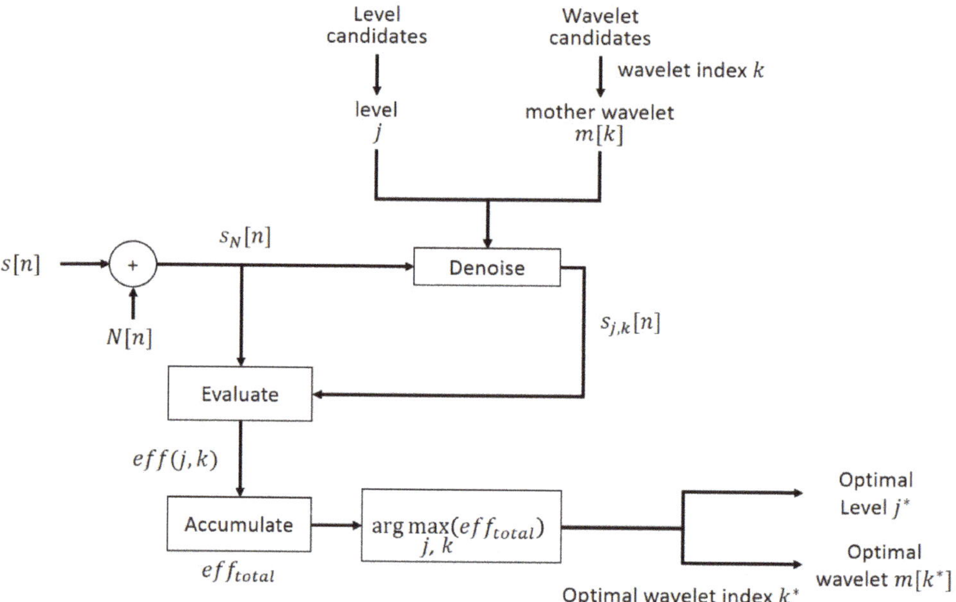

Figure 5. The block diagram of the optimal selection process for the mother wavelet function $m[k^*]$ and decomposition level j^*.

4. Result

4.1. Wavelet Decomposition Level Prediction

To determine the optimal mother wavelet function, the arbitrary decomposition level should be estimated. In this study, the decomposition level is estimated based on the dominant frequency range of the original signal. At the data analysis step, the original DCG signal has the dominant frequency band at 0 to 5 Hz. The approximation coefficient of the decomposition level seven obtain a frequency band of about 0 to 4 Hz, depending on the frequency decomposition rule [68]. Tables 3–8 show the evaluation criterion $eff(j, k)$ values for the candidates of the wavelet families at decomposition levels 3 to 8. According to Tables 3–8, it can be shown that each wavelet family has the maximum $eff(j, k)$ at seven. Thus, the estimated optimal decomposition level is set to seven.

Table 3. The $eff(j, k)$ value of the Coiflets family (wavelet length 18–30).

	3	4	5	6	7	8
18	3.5903	4.0369	4.4724	4.9091	5.0815	3.7178
24	3.5252	3.9538	4.3936	4.8243	5.1840	3.6898
30	3.4757	3.8981	4.3334	4.7615	5.1429	3.6670

Table 4. The $eff(j, k)$ value of the Daubechies family (wavelet length 18–22).

	3	4	5	6	7	8
18	3.5902	4.0328	4.4719	4.9322	5.2873	3.7658
20	3.5671	4.0081	4.4414	4.8843	5.2838	3.7557
22	3.5439	3.9737	4.4199	4.8498	5.2496	3.7590

Table 5. The $eff(j, k)$ value of the Fejer–Korovkin family (wavelet length 18–22).

	3	4	5	6	7	8
18	3.5910	4.0327	4.4659	4.8763	4.9468	3.8317
22	3.5438	3.9782	4.4147	4.8194	4.9721	3.8463

Table 6. The $eff(j, k)$ value of the Biorthogonal Spline family (wavelet length 18–20).

	3	4	5	6	7	8
18	3.5894	4.0377	4.4704	4.9072	5.1879	3.7663
20	3.5521	3.9645	4.3828	4.7759	4.8445	3.8774

Table 7. The $eff(j, k)$ value of the Reverse Biorthogonal Spline family (wavelet length 18–20).

	3	4	5	6	7	8
18	3.5867	4.0361	4.4688	4.9040	5.1384	3.4258
20	3.5602	3.9910	4.4374	4.8952	4.9568	3.0319

Table 8. The $eff(j, k)$ value of the Symlets family (wavelet length 18–20).

	3	4	5	6	7	8
18	3.5914	4.0267	4.4733	4.9256	5.3099	3.7781
20	3.5657	4.0102	4.4428	4.8765	5.2783	3.7603
22	3.5450	3.9741	4.4176	4.8723	5.2523	3.7297

4.2. Most Efficient Mother Wavelet Selection

In this section, the analytic results of the denoising performance using 115 wavelet functions are presented. The examined wavelet functions are selected from the six wavelet families including Coiflets (coif1-coif5), Daubechies (db1-db45), Fejer–Korovkin (fk4-fk22), Biorthogonal Spline (bior1.1-bior6.8), Reverse Biorthogonal Spline (rbio1.1-rbio6.8), and Symlets (sym2-sym30) [69,70]. The length of the wavelet varies by the number of the mother wavelet function name in range of 2 (db1, coif1, etc.) to 90 (db45), only even numbers. Traditionally, the optimal mother wavelet for noise reduction should satisfy the properties of orthogonality, symmetry, regularity, similarity of the shape with the signal, and so on [29,49,71]. According to the experiment result, however, the wavelet length of the mother wavelet function was the key determinant for the noise elimination of the DCG signal. The blue plot of the Figure 6 shows that the $SNR_{j, k}$ of the denoised signal increases with the wavelet length of the mother wavelet function and is saturated at a wavelet length of 18. Although the values of the $SNR_{j, k}$ from the wavelet length of 22 to 90 are similar, execution complexity keeping up with the wavelet length makes the longer wavelet function excessive. As the evaluation criterion $eff(j, k)$ includes both the denoising performance $SNR_{j,\,k}$ and the execution complexity $\eta(k)$, the most efficient wavelet length can be selected as 18 from the red plot of Figure 6. Not only does the execution efficiency become worse but the required length of the signal also increases when the wavelet length grows.

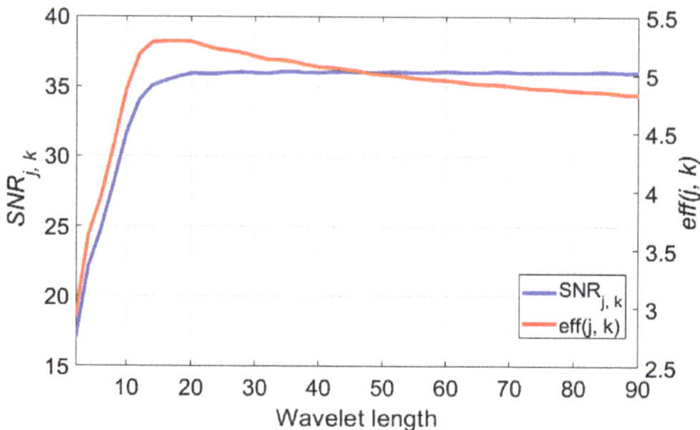

Figure 6. Denoising performance of the wavelet length (2 to 90), blue plot denotes $SNR_{j,\,k}$ value and red plot denotes $eff(j,\,k)$. The used decomposition level j is 7.

However, the wavelet length is not the only feature that represents the characteristic of the mother wavelet function. Though few numbers of the wavelet functions share the same wavelet length, the basis functions and basic features are distinctive. Therefore, even with the same wavelet length, the denoising performance $eff(j,\,k)$ will vary depending on the basis function of the mother wavelet function. At a wavelet length of 16, "db8" and "sym8" scored 5.28 and 5.31 which are higher than other wavelets and the average performance of a wavelet length of 16 is 4.92 (Figure 7a). In the case a wavelet length of 18, "db9" and "sym9" perform better again (Figure 7b). For a wavelet length of 20 (Figure 7c), "db" and "sym" are always the optimal wavelet functions. On the contrary with the Daubechies and Symlets families, the function from the Biorthogonal family shows the lowest performance at all three graphs.

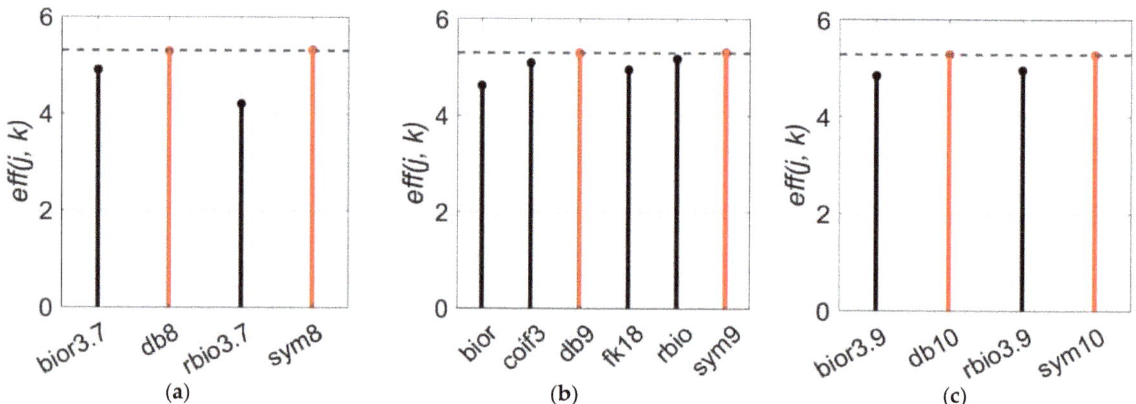

Figure 7. $eff(j,\,k)$ values of six wavelet families (db, sym, coif, fk, bior, and rbio) for wavelet lengths of: (**a**) 16, (**b**) 18, and (**c**) 20. The used decomposition level j is 7.

4.3. Optimal Decomposition Level

To optimize the performance of the denoising method using the wavelet transform and thresholding, determining the proper decomposition level is also necessary. However, the optimal decomposition level was first estimated in Section 4.1 based on the dominant frequency range and the sampling ratio of the signal. In this subsection, we will cover the results of the quantitative analysis for the optimal decomposition level. Figure 8 represents $eff(j, k)$ of the denoised signal. The decomposition level in range from 1 to 10 is denoted as the x-axis, the $eff(j, k)$ from the optimal mother wavelet "db9" is indicated as the y-axis. Figure 9 shows the $eff(j, k)$ of the six wavelet families at decomposition level 1 to 10. Each wavelet function is fixed to a wavelet length of 18, which is the most efficient wavelet length obtained in the previous subsection. Through the experimental result analysis of Figures 8 and 9, the denoising process with decomposition level seven achieves the best performance among all candidates but a higher decomposition level apparently gives less $eff(j, k)$ performance. Although Figure 9 shows that the "bior" graph has a maximum value at level six, still the maximum $eff(j, k)$ of the "bior" family is far less than the maximum $eff(j, k)$ of other families, the wavelet functions from the "bior" family are out of consideration in this study. This incident happens when the wavelet length of the mother wavelet function is too short for the length of the input signal.

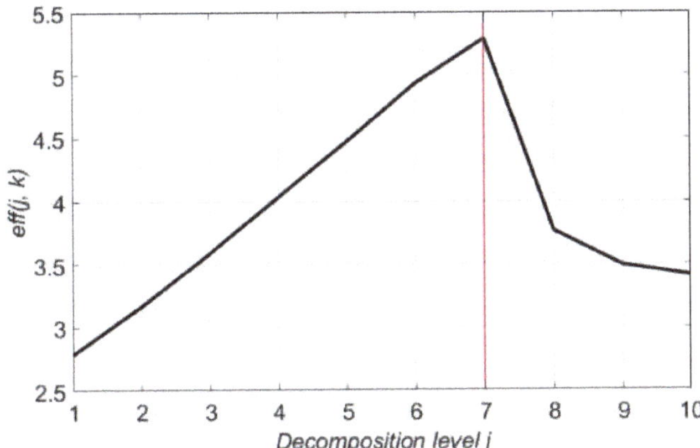

Figure 8. $eff(j, k)$ values of 10 decomposition levels (1 to 10) for db9.

The approximation coefficients at the optimal decomposition level is cA_7, which are composed of a partial signal whose frequency range is approximately 0~4 Hz. This frequency range is the subset of the dominant frequency band (0~5 Hz) of the original signal. Therefore, the decomposition level, which can extract the dominant frequency band of the original signal at approximation coefficients, is the optimal decomposition level for signal denoising. To prove this, the same denoising process was performed to the DCG signals with different sampling ratios. The DCG signals with a sampling frequency of 1000 Hz were sampled to different sampling ratios to provide identical DCG characteristics but different frequency features. The $eff(j, k)$ values of these sampled signals with four different sampling ratios (500 Hz, 250 Hz, 125 Hz, and 62 Hz) are described in Figure 10. As shown in Figure 10a, in case of a sampling ratio of 500 Hz, the most effective decomposition level lowered 1 step to level 6 as the sampling ratio decreased to half of the original. At this time, the denoising performance also decreased. Likewise, the plot of the signal with the sampling frequency of 125 Hz in Figure 10c represents that the optimal decomposition level is four, which is three levels lower than the original. Moreover, the maximum $eff(j, k)$ of the denoised signal decreases as the sampling ratio falls off (Figure 11). To summarize, the optimal decomposition level is determined by the dominant frequency band and the sampling ratio of the original signal. It is also shown that the sampling frequency of the signal affects to the denoising performance because the higher sampling rate can obtain more useful information components than the signal with a lower sampling rate.

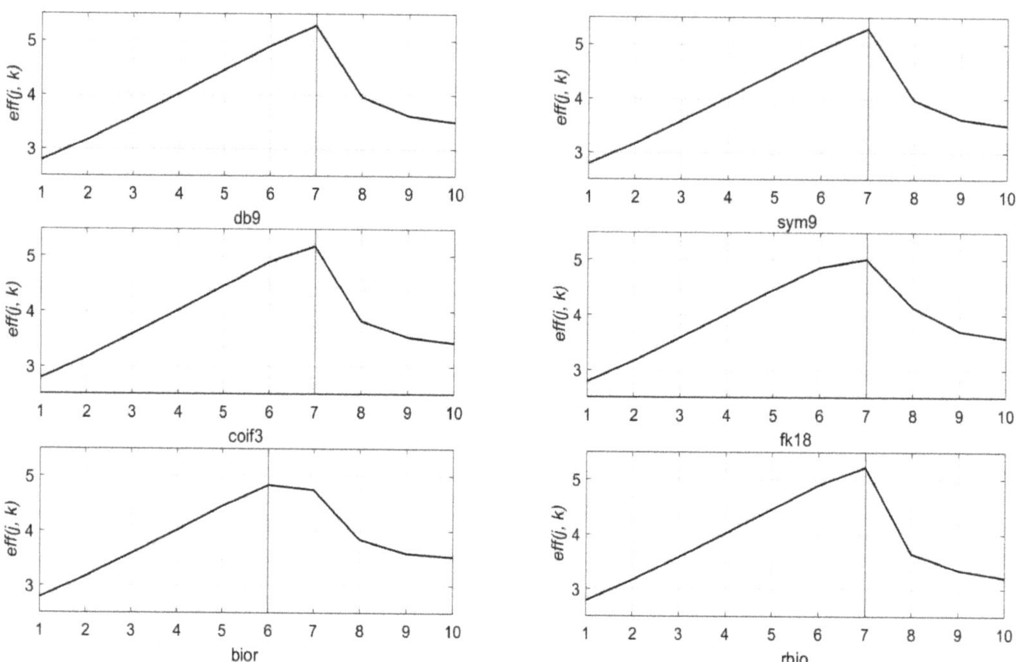

Figure 9. $eff(j, k)$ values of 10 decomposition levels (1 to 10) for six mother wavelet functions (db9, sym9, coif3, fk18, bior, and rbio).

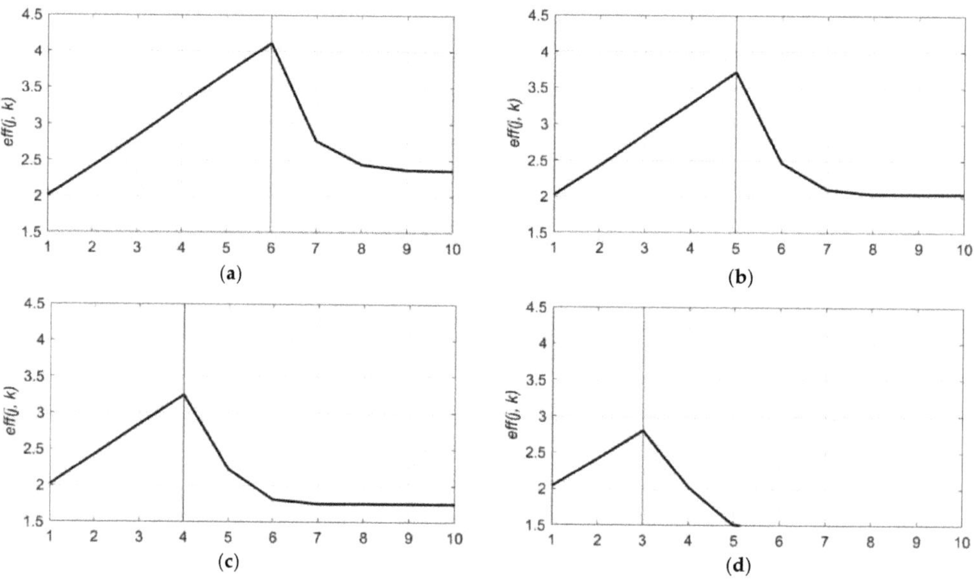

Figure 10. $eff(j, k)$ values of 10 decomposition levels (1 to 10) for db9. (**a**) $eff(j, k)$ values for a sampling frequency of 500 Hz; (**b**) $eff(j, k)$ value for a sampling frequency of 250 Hz; (**c**) $eff(j, k)$ values for a sampling frequency of 125 Hz; and (**d**) $eff(j, k)$ values for a sampling frequency of 62 Hz.

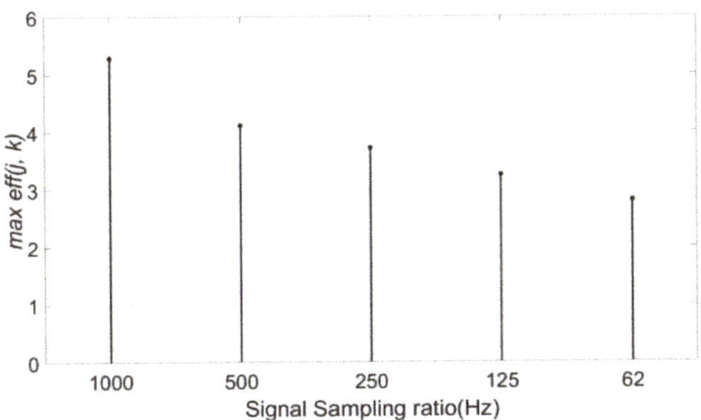

Figure 11. Maximum $eff(j, k)$ values of signals with five sampling ratios (1000 Hz, 500 Hz, 250 Hz, 125 Hz, and 62 Hz).

5. Discussions

In this study, the denoising process using wavelet decomposition and thresholding method is performed to improve the quality of the DCG signal. Although there are many studies of the denoising process using the wavelet function [35,72–74], to the best of the author's knowledge, previous studies have yet to provide the optimal mother wavelet selection for the DCG signal. The purpose of this study is proposing the optimal selection of the mother wavelet function and the decomposition level for the DCG signal to optimize the performance of the denoising process.

Xu et al. [35] and Srivastava et al. [36] proposed the selection of the mother wavelet function and the decomposition level of the object radar signal, respectively. However the proposed mother wavelet function and decomposition level for their object radar signal performed less powerfully on the DCG signal than the proposed selection from this study (Table 9). As a consequence, the optimal mother wavelet function and the decomposition level should be selected identically by analyzing the characteristics of the object signal.

Table 9. The table for the comparison of the $SNR_{j,\,k}$ performance between the Xu et al., Srivastava et al., and proposed selection.

	Proposed Selection	Xu et al. Selection [35]	Srivastava et al. Selection [36]
$SNR_{j,k}$/dB	35.726	27.191	33.153
Wavelet	db9	db4	coif 3
level	7	4	6

The denoising process with the optimal selection for the DCG signal enhanced the quality of the DCG signals in the information field. As the DCG signals can be obtained in a more flexible condition than the ECG signals, the improvement of the DCG signals could enhance the diagnosis of the CVDs.

6. Conclusions

The wavelet decomposition thresholding is a powerful denoising method. The performance of the denoising method can be optimized by using the optimal set of the mother wavelet function and the decomposition level. For the DCG signal, this study suggested the optimal selection of the mother wavelet function and the decomposition level based on the signal analysis. To select the optimal mother wavelet function, the wavelet length of the mother wavelet function is an important element. In this study, the length of the examined signal was 160,000, with a 1000 Hz sampling rate signal recorded for 160 s and the most efficient wavelet length was 18. There are six wavelet families with wavelet length

18; "db9", "sym9", "coif3", "fk18", "bior", and "rbio". Most of these six functions recorded superior performance to other functions with a different wavelet length and "db9" and "sym9" were selected for the optimal mother wavelet functions among all 115 wavelet candidates. The optimal decomposition level for the DCG signal was determined as seven, which shows that the estimation of the decomposition level based on the length of the signal was precise.

Since the noise reduction method based on the wavelet decomposition and the thresholding with optimal parameters successfully removes the noise from the DCG signal, the quality of the heart rate signal obtained by Doppler radar sensors was improved in the information field. Therefore, analyzing a DCG signal using artificial neural networks for the diagnosis of CVDs is conducted as the aim of the future work following to this study.

The major findings in this study is denoted as follows:

- The wavelet length of the mother wavelet function was the important element for the selection of the most efficient mother wavelet. The longer mother wavelet function did not provide a better denoising performance. As the longer wavelet function requires more performance complexity, the optimal wavelet length for the performance efficiency should be considered;
- The optimal decomposition level was determined by the sampling frequency and dominant frequency range of the original signal. The level that could decompose the dominant frequency range from the signal was the optimal decomposition level. For this reason, the optimal decomposition level could be predicted based on the signal analysis in the frequency domain. The appropriate decomposition level produced a modest threshold value for noise removal;
- The higher the sampling frequency of the DCG signal, the more powerful the performance of the denoising process. The higher sampling frequency enabled the signal to obtain more useful components.

Author Contributions: Conceptualization, Y.I.J., J.-R.Y. and N.K.K.; Data curation, J.Y.S.; Formal analysis, Y.I.J. and N.K.K.; Funding acquisition, J.-R.Y. and N.K.K.; Investigation, Y.I.J.; Methodology, Y.I.J. and N.K.K.; Project administration, J.-R.Y. and N.K.K.; Resources, J.Y.S.; Software, Y.I.J. and J.Y.S.; Supervision, J.-R.Y. and N.K.K.; Validation, Y.I.J.; Visualization, Y.I.J.; Writing—original draft, Y.I.J.; Writing—review & editing, N.K.K. All authors have read and agreed to the published version of the manuscript.

Funding: This work was supported in part by the National Research Foundation of Korea (NRF) through the Basic Science Research Program funded by the Ministry of Education under Grant 2020R1F1A1064330, and in part by Yeungnam University, through a research grant, in 2019.

Institutional Review Board Statement: Not applicable.

Informed Consent Statement: Informed consent was obtained from all subjects involved in the study.

Data Availability Statement: The data are not publicly available due to the company security policy and personal protection of subjects.

Acknowledgments: The authors wish to represent proper appreciations to all researchers who presented their studies and efforts for this study.

Conflicts of Interest: The authors declare no conflict of interest.

References

1. Wang, H.; Naghavi, M.; Allen, C.; Barber, R.M.; Bhutta, Z.A.; Casey, D.C.; Charlson, F.J.; Chen, A.Z.; Coates, M.M.; Coggeshall, M.; et al. Global, regional, and national life expectancy, all-cause mortality, and cause-specific mortality for 249 causes of death, 1980–2015: A systematic analysis for the Global Burden of Disease Study 2015. *Lancet* **2016**, *388*, 1459–1544. [CrossRef]
2. Deng, M.; Wang, C.; Tang, M.; Zheng, T. Extracting cardiac dynamics within ECG signal for human identification and cardiovascular diseases classification. *Neural Netw.* **2018**, *100*, 70–83. [CrossRef] [PubMed]
3. Nemati, E.; Deen, M.J.; Mondal, T. A wireless wearable ECG sensor for long-term applications. *IEEE Commun. Mag.* **2012**, *50*, 36–43. [CrossRef]

Pani, D.; Achilli, A.; Bassareo, P.P.; Cugusi, L.; Mercuro, G.; Fraboni, B.; Bonfiglio, A. Fully: Textile polymer: Based ECG electrodes: Overcoming the limits of metal: Based textiles. In Proceedings of the 2016 Computing in Cardiology Conference (CinC); Computing in Cardiology, Vancouver, BC, Canada, 11–14 September 2016; pp. 373–376.

Castro, I.D.; Mercuri, M.; Torfs, T.; Lorato, I.; Puers, R.; Van Hoof, C. Sensor Fusion of Capacitively Coupled ECG and Continuous-Wave Doppler Radar for Improved Unobtrusive Heart Rate Measurements. *IEEE J. Emerg. Sel. Top. Circuits Syst.* **2018**, *8*, 316–328. [CrossRef]

Lu, L.; Li, C.; Lie, D.Y.C. Experimental demonstration of noncontact pulse wave velocity monitoring using multiple Doppler radar sensors. In Proceedings of the 2010 Annual International Conference of the IEEE Engineering in Medicine and Biology, Piscataway, NJ, USA, 31 August–4 September 2010; Volume 2010, pp. 5010–5013.

Mogi, E.; Ohtsuki, T. Heartbeat detection with Doppler sensor using adaptive scale factor selection on learning. In Proceedings of the 2015 IEEE 26th Annual International Symposium on Personal, Indoor, and Mobile Radio Communications (PIMRC), Hong Kong, China, 30 August–2 September 2015; pp. 2166–2170.

Choi, C.; Park, J.; Lee, H.; Yang, J. Heartbeat detection using a Doppler radar sensor based on the scaling function of wavelet transform. *Microw. Opt. Technol. Lett.* **2019**, *61*, 1792–1796. [CrossRef]

Hu, W.; Zhao, Z.; Wang, Y.; Zhang, H.; Lin, F. Noncontact Accurate Measurement of Cardiopulmonary Activity Using a Compact Quadrature Doppler Radar Sensor. *IEEE Trans. Biomed. Eng.* **2013**, *61*, 725–735. [CrossRef]

Du, L.; Wang, B.; Wang, P.; Ma, Y.; Liu, H. Noise Reduction Method Based on Principal Component Analysis With Beta Process for Micro-Doppler Radar Signatures. *IEEE J. Sel. Top. Appl. Earth Obs. Remote. Sens.* **2015**, *8*, 4028–4040. [CrossRef]

Lee, Y.S.; Pathirana, P.N.; Steinfort, C.L.; Caelli, T. Monitoring and Analysis of Respiratory Patterns Using Microwave Doppler Radar. *IEEE J. Transl. Eng. Health Med.* **2014**, *2*, 1–12. [CrossRef]

Li, C.; Lubecke, V.M.; Boric-Lubecke, O.; Lin, J. A Review on Recent Advances in Doppler Radar Sensors for Noncontact Healthcare Monitoring. *IEEE Trans. Microw. Theory Tech.* **2013**, *61*, 2046–2060. [CrossRef]

Mogi, E.; Ohtsuki, T. Heartbeat detection with Doppler radar based on spectrogram. In Proceedings of the 2017 IEEE International Conference on Communications (ICC), Paris, France, 21–25 May 2017; pp. 1–6.

Rahman, A.; Yavari, E.; Gao, X.; Lubecke, V.; Boric-Lubecke, O. Signal processing techniques for vital sign monitoring using mobile short range doppler radar. In Proceedings of the 2015 IEEE Topical Conference on Biomedical Wireless Technologies, Networks, and Sensing Systems (BioWireleSS), San Diego, CA, USA, 25–28 January 2015; pp. 1–3.

Chervyakov, N.; Lyakhov, P.; Kaplun, D.; Butusov, D.; Nagornov, N. Analysis of the Quantization Noise in Discrete Wavelet Transform Filters for Image Processing. *Electronics* **2018**, *7*, 135. [CrossRef]

Antoniadis, A.; Oppenheim, G. *Wavelets and Statistics*; Springer Science & Business Media: Berlin/Heidelberg, Germany, 2012; Volume 103.

Kovacevic, J.; Goyal, V.K.; Vetterli, M. Fourier and wavelet signal processing. *Fourier Wavelets.org*. 2013, pp. 1–294. Available online: http://www.fourierandwavelets.org/FWSP_a3.2_2013.pdf (accessed on 3 March 2021).

Rafiee, J.; Rafiee, M.; Prause, N.; Schoen, M. Wavelet basis functions in biomedical signal processing. *Expert Syst. Appl.* **2011**, *38*, 6190–6201. [CrossRef]

Young, R.K. *Wavelet Theory and Its Applications*; Springer International Publishing: Berlin/Heidelberg, Germany, 1993; Volume 189.

Rai, H.M.; Trivedi, A.; Shukla, S. ECG signal processing for abnormalities detection using multi-resolution wavelet transform and Artificial Neural Network classifier. *Measurement* **2013**, *46*, 3238–3246. [CrossRef]

Krishnaveni, V.; Jayaraman, S.; Aravind, S.; Hariharasudhan, V.; Ramadoss, K. Automatic Identification and Removal of Ocular Artifacts from EEG using Wavelet Transform. *Meas. Sci. Rev.* **2006**, *6*, 45–57.

Torrence, C.; Compo, G.P. A practical guide to wavelet analysis. *Bull. Am. Meteorol. Soc.* **1998**, *79*, 61–78. [CrossRef]

Percival, D.B.; Walden, A.T. *Wavelet Methods for Time Series Analysis*; Cambridge University Press: Cambridge, UK, 2000; Volume 4.

Jansen, M.; Bultheel, A. Asymptotic behavior of the minimum mean squared error threshold for noisy wavelet coefficients of piecewise smooth signals. *IEEE Trans. Signal Process.* **2001**, *49*, 1113–1118. [CrossRef]

Jansen, M.M. Minimum risk thresholds for data with heavy noise. *IEEE Signal Process. Lett.* **2006**, *13*, 296–299. [CrossRef]

Al-Qazzaz, N.K.; Ali, S.H.B.M.; Ahmad, S.A.; Islam, M.S.; Escudero, J. Selection of Mother Wavelet Functions for Multi-Channel EEG Signal Analysis during a Working Memory Task. *Sensors* **2015**, *15*, 29015–29035. [CrossRef] [PubMed]

Sang, Y.-F.; Wang, D.; Wu, J.-C. Entropy-Based Method of Choosing the Decomposition Level in Wavelet Threshold De-noising. *Entropy* **2010**, *12*, 1499–1513. [CrossRef]

Engin, E.Z.; Arslan, Ö. Selection of Optimum Mother Wavelet Function for Turkish Phonemes. *Int. J. Appl. Math. Electron. Comput.* **2019**, *7*, 56–64. [CrossRef]

Bhatia, P.; Boudy, J.; Andreão, R. Wavelet transformation and pre-selection of mother wavelets for ECG signal processing. In Proceedings of the 24th IASTED International Conference on Biomedical Engineering, Innsbruck, Austria, 15–17 February 2006; pp. 390–395.

Castillo, E.; Morales, D.P.; García, A.; Martínez-Martí, F.; Parrilla, L.; Palma, A.J. Noise Suppression in ECG Signals through Efficient One-Step Wavelet Processing Techniques. *J. Appl. Math.* **2013**, *2013*, 1–13. [CrossRef]

Li, F.; Meng, G.; Kageyama, K.; Su, Z.; Ye, L. Optimal Mother Wavelet Selection for Lamb Wave Analyses. *J. Intell. Mater. Syst. Struct.* **2009**, *20*, 1147–1161. [CrossRef]

32. Dan, G.; Li, Z.; Ding, H. A Mother Wavelet Selection Algorithm for Respiratory Rate Estimation from Photoplethysmogram. In Proceedings of the 26th Brazilian Congress on Biomedical Engineering, Vitoria, Brazil, 26–30 October 2015; Springer International Publishing: Berlin/Heidelberg, Germany, 2015; Volume 51, pp. 962–965.

33. Saraswathy, J.; Hariharan, M.; Nadarajaw, T.; Khairunizam, W.; Yaacob, S. Optimal selection of mother wavelet for accurate infant cry classification. *Australas. Phys. Eng. Sci. Med.* **2014**, *37*, 439–456. [CrossRef]

34. Strömbergsson, D.; Marklund, P.; Berglund, K.; Saari, J.; Thomson, A. Mother wavelet selection in the discrete wavelet transform for condition monitoring of wind turbine drivetrain bearings. *Wind. Energy* **2019**, *22*, 1581–1592. [CrossRef]

35. Xu, X.; Luo, M.; Tan, Z.; Pei, R. Echo signal extraction method of laser radar based on improved singular value decomposition and wavelet threshold denoising. *Infrared Phys. Technol.* **2018**, *92*, 327–335. [CrossRef]

36. Srivastava, M.; Anderson, C.L.; Freed, J.H. A New Wavelet Denoising Method for Selecting Decomposition Levels and Noise Thresholds. *IEEE Access* **2016**, *4*, 3862–3877. [CrossRef] [PubMed]

37. Adeli, H.; Zhou, Z.; Dadmehr, N. Analysis of EEG records in an epileptic patient using wavelet transform. *J. Neurosci. Methods* **2003**, *123*, 69–87. [CrossRef]

38. Mallat, S. A theory for multiresolution signal decomposition: The wavelet representation. *IEEE Trans. Pattern Anal. Mach. Intell.* **1989**, *11*, 674–693. [CrossRef]

39. Soman, K.; Ramachandran, K. *Insight into Wavelets: From Theory to Practice*; PHI Learning Ltd.: New Delhi, India, 2010.

40. Parameswariah, C.; Cox, M. Frequency characteristics of wavelets. *IEEE Trans. Power Deliv.* **2002**, *17*, 800–804. [CrossRef]

41. Sanei, S.; Chambers, J.A. *EEG Signal Processing*; John Wiley & Sons: Toronto, ON, Canada, 2013.

42. Rao, R. Wavelet Transforms. *Encycl. Imaging Sci. Technol.* **2002**. [CrossRef]

43. Park, J.-H.; Yang, J.-R. Multiphase Continuous-Wave Doppler Radar With Multiarc Circle Fitting Algorithm for Small Periodic Displacement Measurement. *IEEE Trans. Microw. Theory Tech.* **2020**, *1*. [CrossRef]

44. Tan, H.R.; Tan, A.; Khong, P.; Mok, V. Best Wavelet Function Identification System for ECG signal denoise applications. In Proceedings of the 2007 International Conference on Intelligent and Advanced Systems, Kuala Lumpur, Malaysia, 25–28 November 2007; pp. 631–634.

45. Stantic, D.; Jo, J. Selection of Optimal Parameters for ECG Signal Smoothing and Baseline Drift Removal. *Comput. Inf. Sci.* **2014**, *99*. [CrossRef]

46. Bangerter, N.K.; Hargreaves, B.A.; Vasanawala, S.S.; Pauly, J.M.; Gold, G.E.; Nishimura, D.G. Analysis of multiple-acquisition SSFP. *Magn. Reson. Med.* **2004**, *51*, 1038–1047. [CrossRef]

47. Ali, M.N.; El-Dahshan, E.-S.A.; Yahia, A.H. Denoising of Heart Sound Signals Using Discrete Wavelet Transform. *Circuits Syst. Signal. Process.* **2017**, *36*, 4482–4497. [CrossRef]

48. Jondral, F.K. White Gaussian Noise—Models for Engineers. *Frequenz* **2017**, *72*, 293–299. [CrossRef]

49. He, H.; Tan, Y.; Wang, Y. Optimal Base Wavelet Selection for ECG Noise Reduction Using a Comprehensive Entropy Criterion. *Entropy* **2015**, *17*, 6093–6109. [CrossRef]

50. Boyat, A.K.; Joshi, B.K. A Review Paper: Noise Models in Digital Image Processing. *Signal. Image Process. Int. J.* **2015**, *6*, 63–75. [CrossRef]

51. Donoho, D. De-noising by soft-thresholding. *IEEE Trans. Inf. Theory* **1995**, *41*, 613–627. [CrossRef]

52. Elshorbagy, A.; Simonovic, S.; Panu, U. Noise reduction in chaotic hydrologic time series: Facts and doubts. *J. Hydrol.* **2002**, *256*, 147–165. [CrossRef]

53. Natarajan, B. Filtering random noise from deterministic signals via data compression. *IEEE Trans. Signal. Process.* **1995**, *43*, 2595–2605. [CrossRef]

54. Addison, P.S. Wavelet transforms and the ECG: A review. *Physiol. Meas.* **2005**, *26*, R155–R199. [CrossRef]

55. Gacek, A.; Pedrycz, W. *ECG Signal Processing, Classification and Interpretation: A Comprehensive Framework of Computational Intelligence*; Springer Science & Business Media: Berlin/Heidelberg, Germany, 2011.

56. Stéphane, M. *A Wavelet Tour of Signal Processing*; Elsevier: San Diego, CA, USA, 2009. [CrossRef]

57. Johnstone, I.M.; Ilverman, B.W.S. Needles and straw in haystacks: Empirical Bayes estimates of possibly sparse sequences. *Ann. Stat.* **2004**, *32*, 1594–1649. [CrossRef]

58. Cai, T.T. On block thresholding in wavelet regression: Adaptivity, block size, and threshold level. *Stat. Sin.* **2002**, *12*, 1241–1273.

59. Donoho, D.L.; Johnstone, I.M. Adapting to unknown smoothness via wavelet shrinkage. *J. Am. Stat. Assoc.* **1995**, *90*, 1200–1224. [CrossRef]

60. Donoho, D.L.; Johnstone, I.M. Minimax estimation via wavelet shrinkage. *Ann. Stat.* **1998**, *26*, 879–921. [CrossRef]

61. Zhang, X.-P.; Desai, M.D. Adaptive denoising based on SURE risk. *IEEE Signal. Process. Lett.* **1998**, *5*, 265–267. [CrossRef]

62. Poornachandra, S. Wavelet-based denoising using subband dependent threshold for ECG signals. *Digit. Signal. Process.* **2008**, *18*, 49–55. [CrossRef]

63. Reddy, G.U.; Muralidhar, M.; Varadarajan, S. ECG De-Noising using improved thresholding based on Wavelet transforms. *Int. J. Comput. Sci. Netw. Secur.* **2009**, *9*, 221–225.

64. He, C.; Xing, J.; Li, J.; Yang, Q.; Wang, R. A New Wavelet Threshold Determination Method Considering Interscale Correlation in Signal Denoising. *Math. Probl. Eng.* **2015**, *2015*, 1–9. [CrossRef]

65. Poornachandra, S.; Kumaravel, N. A novel method for the elimination of power line frequency in ECG signal using hyper shrinkage function. *Digit. Signal. Process.* **2008**, *18*, 116–126. [CrossRef]

66. Huimin, C.; Ruimei, Z.; Yanli, H. Improved Threshold Denoising Method Based on Wavelet Transform. *Phys. Procedia* **2012**, *33*, 1354–1359. [CrossRef]

7. Donoho, D.L.; Johnstone, J.M. Ideal spatial adaptation by wavelet shrinkage. *Biometrika* **1994**, *81*, 425–455. [CrossRef]
8. Bilgin, S.; Colak, O.H.; Koklukaya, E.; Arı, N.; Ari, N. Efficient solution for frequency band decomposition problem using wavelet packet in HRV. *Digit. Signal. Process.* **2008**, *18*, 892–899. [CrossRef]
9. Rafiee, J.; Schoen, M.; Prause, N.; Urfer, A.; Rafiee, M. A comparison of forearm EMG and psychophysical EEG signals using statistical signal processing. In Proceedings of the 2009 2nd International Conference on Computer, Control and Communication, Karachi, Pakistan, 17–18 February 2009; pp. 1–5.
10. Khanam, R.; Ahmad, S.N. Selection of wavelets for evaluating SNR, PRD and CR of ECG signal. *Int. J. Eng. Sci. Innov. Technol* **2013**, *2*, 112–119.
11. Ngui, W.K.; Leong, M.S.; Hee, L.M.; Abdelrhman, A.M. Wavelet Analysis: Mother Wavelet Selection Methods. *Appl. Mech. Mater.* **2013**, *393*, 953–958. [CrossRef]
12. Thatiparthi, S.; Gudheti, R.; Sourirajan, V. MST Radar Signal Processing Using Wavelet-Based Denoising. *IEEE Geosci. Remote. Sens. Lett.* **2009**, *6*, 752–756. [CrossRef]
13. Rock, J.; Toth, M.; Messner, E.; Meissner, P.; Pernkopf, F. Complex signal denoising and interference mitigation for automotive radar using convolutional neural networks. In Proceedings of the 2019 22th International Conference on Information Fusion (FUSION), Ottawa, ON, Canada, 2–5 July 2019; pp. 1–8.
14. Hua, T.; Dai, K.; Zhang, X.; Yao, Z.; Wang, H.; Xie, K.; Feng, T.; Zhang, H. Optimal VMD-Based Signal Denoising for Laser Radar via Hausdorff Distance and Wavelet Transform. *IEEE Access* **2019**, *7*, 167997–168010. [CrossRef]

Article

Machine Learning-Based Human Recognition Scheme Using a Doppler Radar Sensor for In-Vehicle Applications

Eugin Hyun [1,*] , **Young-Seok Jin** [1], **Jae-Hyun Park** [2] and **Jong-Ryul Yang** [2]

[1] Division of Automotive Technology, ICT Research Institute, Convergence Research Institute, DGIST, 333 Techno Jungang-daero, Hyeonpung-myeon, Dalseong-gun, Daegu 42988, Korea; ysjin@dgist.ac.kr

[2] Department of Electronic Engineering, Yeungnam University, Gyeongsan, Gyeongbuk-do 38541, Korea; bravopark@ynu.ac.kr (J.-H.P.); jryang@yu.ac.kr (J.-R.Y.)

* Correspondence: braham@dgist.ac.kr; Tel.: +82-53-785-4560

Received: 29 September 2020; Accepted: 29 October 2020; Published: 30 October 2020

Abstract: In this paper, we propose a Doppler spectrum-based passenger detection scheme for a CW (Continuous Wave) radar sensor in vehicle applications. First, we design two new features, referred to as an 'extended degree of scattering points' and a 'different degree of scattering points' to represent the characteristics of the non-rigid motion of a moving human in a vehicle. We also design one newly defined feature referred to as the 'presence of vital signs', which is related to extracting the Doppler frequency of chest movements due to breathing. Additionally, we use a BDT (Binary Decision Tree) for machine learning during the training and test steps with these three extracted features. We used a 2.45 GHz CW radar front-end module with a single receive antenna and a real-time data acquisition module. Moreover, we built a test-bed with a structure similar to that of an actual vehicle interior. With the test-bed, we measured radar signals in various scenarios. We then repeatedly assessed the classification accuracy and classification error rate using the proposed algorithm with the BDT. We found an average classification accuracy rate of 98.6% for a human with or without motion.

Keywords: passenger detection; CW radar; radar feature vector; radar machine learning

---

1. Introduction

The fact that children are dying in hot vehicles has recently become a major social issue. Thus, the European NCAP (New Car Evaluation Program) has recommended the installation of CPD (Child Presence Detection) technology on all new cars starting in 2020 [1]. Moreover, many countries' safety regulators have also considered rules that could mandate CPD systems aimed to detect a child left in a vehicle. To support such a system, various sensors capable of detecting objects in vehicles or monitoring vehicle body status are required [2,3].

Another application of passenger detection is in electric vehicles. In electric vehicles, the heating and air conditioning functions depend on the efficiency of the battery [4]. If heating and cooling systems in vehicles can be automatically controlled for each seat, battery consumption can be decreased. Thus, to support these functions, technology to detect passengers in each seat is required.

One additional application for occupant recognition is in self-driving vehicles. The driving operation for an autonomous vehicle strongly depends on the presence or absence of passengers. That is, when occupants are riding, the comfort and reliability of passengers become very important issues [5]. Moreover, depending on whether the occupant is sleeping or moving, the self-driving style can differ. Thus, it is very important to assess the occupancy and status of passengers in every seat.

For the various applications described above, the performances capabilities of sensors to detect passenger are very important. The characteristics of these sensors are described below.

Among various methods, one very simple approach is to use a pressure sensor. However, when any object is placed on a seat, it is impossible to determine whether or not the object is a human. Another solution is to measure and use the distance from the object by means of an ultrasonic sensor. However, this method also cannot distinguish between different types of objects.

Recently, thermal infrared sensors have been attracting attention given their ability to check for the presence of passengers using human temperature. However, this method is highly sensitive to a person's clothing or to the external temperature condition.

The use of a camera is also a very effective solution for these applications. Specifically, because stereo cameras and depth cameras can measure the distance to an object, they can recognize various motions of a human when applying deep learning with image features. However, camera sensors are limited due to the external lighting conditions. Another disadvantage is that the amount of computation for image processing is excessive. Moreover, the installation of a camera inside their vehicles may cause consumers to reject these vehicles due to privacy issues.

Recently, a radar sensor-based occupant detection system has attracted attention, as radar is robust to external conditions [6]. Moreover, radar can distinguish between a moving object and a stationary object, and these systems can also monitor vital signal of sleeping or non-moving humans in a vehicle [7].

To detect the motion of an object and to detect human vital signs using a radar sensor, popular types currently in use are impulse UWB (Ultra Wide Band) radar, FMCW (Frequency Modulated Continuous Wave) radar, and CW (Continuous Wave) radar.

In impulse UWB radar, because it is possible to measure a high-resolution range, we can distinguish chest movements due to breathing by measuring changes in distance values. Because high-resolution range detection is most advantageous, UWB radar is widely used for vital sign recognition [8,9]. Earlier work proposed the concept of detecting the vital signals of a passenger by mounting a UWB radar sensor in a vehicle [8]. In another approach [9], a UWB radar sensor was used to detect human vital signs for each seat, applying the features extracted from the detected range into the machine leaning approach, such as a SVM (Support Vector Machine). However, these two related works only focused on non-moving humans, and did not consider moving humans or other objects.

Although UWB radar is very popular, the Doppler component cannot be detected in order to distinguish between a stationary object and a moving object. Thus, an additional algorithm is required using measured distances. In addition, because this type transmits an impulse-shaped waveform in the time domain, high peak transmission is limited. This can result in a low SNR (Signal-to-Noise Ratio) over a certain distance.

Recently, because FMCW radar can measure both the distance and the Doppler information, FMCW radar has come to be commonly used in commercial applications. Moreover, when detecting changes in phases over several periods, the respiration period was extracted in earlier works [10,11]. In addition, one related study [10] presented a method that separated the vital signs reflected from two humans using a high-resolution algorithm, in that case the MUSIC (Multiple Signal Classifier) algorithm. However, neither method focused on only vital sign signal detection, nor were they intended for in-vehicle applications.

Despite the fact that the FMCW radar is widely used in commercial applications, a PLL (Phase Loop Lock) circuit is also required to synchronize the transmit waveform phases and to ensure linearity during the modulation step. Moreover, because FMCW radar can detect moving and stationary objects, extra algorithms to distinguish them are necessary.

Finally, the CW radar sensor is a popular radar sensor due to its very simple hardware structure. However, because CW radar can only receive Doppler signals, these sensors can only detect moving objects, and cannot detect the range. Thus, in CW radar, breathing signals can be measured by analyzing the Doppler signal generated from the chest movements [7,12,13]. One earlier study [12] presented the concept of recognizing a human remaining in a vehicle using CW radar by cancelling the background noise. In another work [13], an algorithm for detecting not only respiration, but also the heartbeat, was proposed. Because CW radar is somewhat sensitive to external noise, one study [7] proposed attaching an accelerometer to the radar sensor to record vibrations of the vehicle itself. However, these three related works also considered stationary humans when extracting human vital signs.

Although though CW radar sensors have various disadvantages, they can be easily applied in various applications as a low-cost senor compared to UWB radar and FMCW radar. Thus, in this paper, we employ the CW radar type to realize a passenger detection system with a very simple architecture.

To effectively confirm the existence of a human in a vehicle, we can recognize a human who is still, sleeping, or moving using the Doppler signal measured using the CW radar sensor. That is, we can determine whether or not the detected moving object is a human and can extract vital signals from a non-moving human on the seat.

If a human is moving on a seat, the echo signal of the human's vital signs can be masked by the Doppler signal of the human's motion. In such cases, it is difficult to determine whether a human is present or not in a vehicle using only the detected vital signal.

Moreover, if an inanimate object is moving on a seat or if the vehicle itself has vibration, the radar system should be able to recognize a Doppler signal. Thus, when using only the presence of the Doppler echo, it is impossible to determine whether or not a human is occupying a seat.

Thus, in this paper, we propose a human recognition concept as part of our effort to implement a proper passenger detection system, as shown in Figure 1. The wait mode transmits the received radar signal into the motion-detection mode and the vital sign detection mode, with both modes operating in parallel.

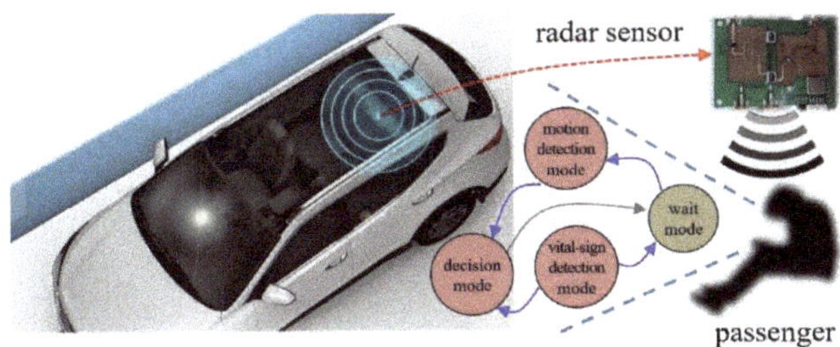

Figure 1. Human recognition concept for in-vehicle application. Here, the car photo on the left, also used in an earlier work [2], was modified somewhat.

Based the results of both modes, in the decision mode, human recognition is determined, and the system reverts back to the wait mode.

For the vital sign detection mode, we design a simple vital sign detection algorithm to determine whether breathing is present or not. Thus, we extract one feature vector to indicate the presence of vital signs.

For the motion detection mode, we propose algorithms to determine whether or not a moving object in a vehicle is a human. In the proposed method, we use the human characteristics of non-rigid motion. That is, in the case of a human, because the radar signal is reflected from various components of the human's torso, head, shoulders, arms, waist, and thighs, micro-Doppler effect appears. Thus, in paper, we initially generate a micro-Doppler image in the time-frequency domain. Next, we design two new feature vectors that suitably represent the characteristics of a moving human in a vehicle from the micro-Doppler image.

Finally, in the decision mode, we conduct machine learning using a BDT (Binary Decision Tree), which has a very simple structure, and the proposed three features to determine the presence of passengers in vehicles.

Thus, we extract three features using actual measurement data from a CW radar transceiver and verify the proposed machine learning-based human recognition scheme.

In Section 2, we present the proposed human recognition scheme with machine learning. In Section 3, we present the verification results using actual data from a 2.45 GHz CW radar front-end module and a real-time data acquisition module. Finally, we present the conclusion of our study and the suggestions for future work in the final Section.

2. Proposed Human Recognition Scheme in a Vehicle

2.1. Problem Definition of Human Detection in a Vehicle

In this paper, Figure 2 shows the information detected via radar sensor according to the increased level of motion of passengers in a vehicle. Here, the x-axis indicates the amount of movement by a human and the y-axis represents the Doppler frequency of the echo signal.

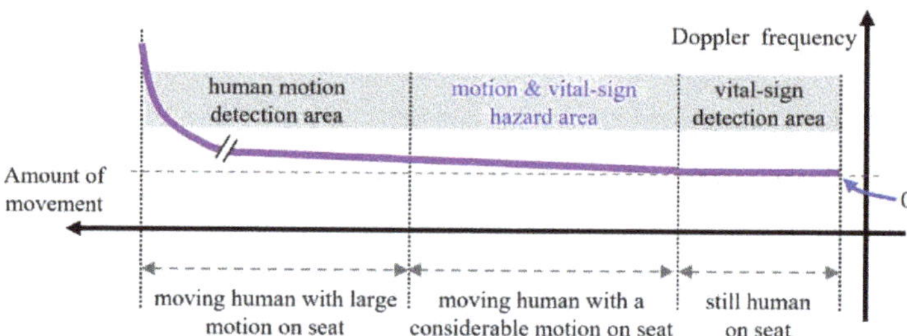

Figure 2. Information detected from a radar sensor according to amount of movement by a human on a seat in the vehicle.

In a vehicle, when a passenger is sleeping or a still human is sitting on a seat, Doppler radar sensors can detect breathing signals from vital sign detection area in Figure 2. Accordingly, we can easily distinguish between a human and an inanimate object using the echo generated by the human's respiration.

However, if a passenger moves with much motion on the seat, as shown in the motion detection area of Figure 2, the Doppler components issued by the body motion can mask most of the weak vital sign signals. In such cases, if an object is moving with considerable motion, we can detect the object using the Doppler echo level. However, with only the Doppler component, it is impossible to confirm whether or not a human has been detected.

Moreover, in the motion and vital sign hazard area, the detection of vital signs depends entirely on the amount of human movement. In other words, when a human is moving with relatively slight motion, motion and vital signals can appear together, whereas the vital signals can be immediately masked when a passenger is moving with a Doppler volume above a certain level.

Therefore, in both the motion detection area and the motion and vital sign hazard area, we cannot distinguish between a human and another object by the presence of vital signals or Doppler signals alone. To overcome this problem, we propose here a human recognition scheme using the characteristics of the echoed Doppler spectra of the object.

Generally, in signals reflected from a walking or running human, sidebands appear around the Doppler frequency due to the non-rigid motion. That is, various Doppler echoes can be extracted as reflected scatters of human components, such as the body, arms, and feet. Moreover, the distribution of Doppler scattering points received from a human can vary greatly during the measurement time [14,15]. In one earlier study [14], based on the FMCW radar, we proposed a concept to distinguish between pedestrians and vehicles on a road by detecting the range and velocity and analyzing the pattern of the Doppler spectrum. In addition, in another study [15], we proposed a classification algorithm for humans and vehicles that extracted feature vectors from the received FMCW radar signal and applied them to machine learning.

From these hints, in vehicle applications, although the received power is weak and there are fewer scattering points compared to the case of a walking human, the received signal has multiple reflection points echoed from the head, torso, waist, arms, pelvis, and other parts of the passenger's body. Moreover, because the human in the vehicle cannot move constantly, the Doppler spectrum will vary more over time.

Thus, we can find certain patterns in the micro-Doppler image of a moving passenger. In this paper, we distinguish between humans and other objects by using feature vectors extracted from this pattern, together with the vital signal extracted through additional signal processing, and applying the features to machine learning.

2.2. Concept of Proposed Human Recognition

Figure 3a presents the proposed concept of recognizing a passenger in a vehicle based on the Doppler spectra and vital signs of the passenger. In Figure 3b, we illustrate an example of the scattering points reflected from the passenger and from a box. In this case, we cannot know which aspects of the human's various components are reflected. Moreover, we cannot know the extent to which individual parts of body contribute to the reflection.

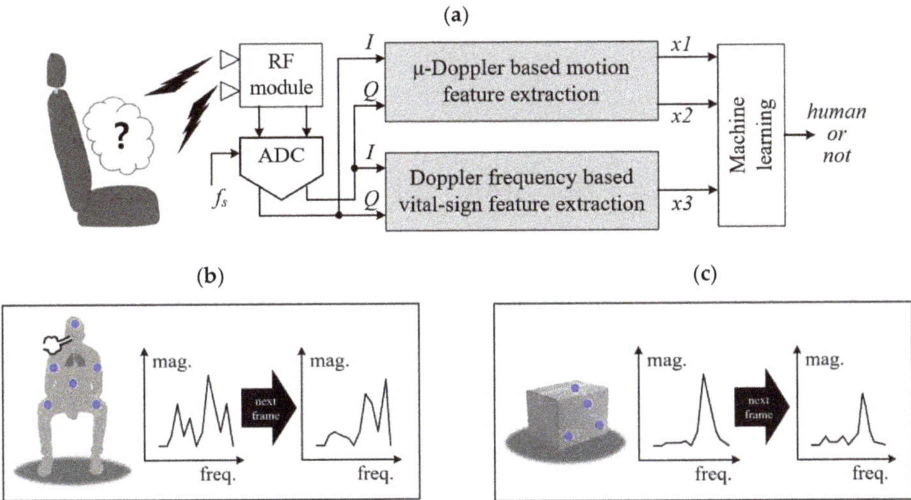

Figure 3. Proposed human recognition scheme together with micro-Doppler and vital signals using machine learning for a Doppler radar sensor: (**a**) top block diagram of the proposed algorithm, (**b**) example of Doppler scattering points of a human, (**c**) example of Doppler scattering points of a non-human and a non-human object.

However, as explained above, if a human is moving, the Doppler spectrum reflected by the human is expected to spread and change over time due non-rigid motion, as shown in Figure 3b. On the other hand, echoes from a non-human object such as a moving box are expected to have a sharp type of Doppler spectrum, as shown in Figure 3c. These characteristics are the motivation behind the algorithm proposed in this paper.

The proposed human recognition scheme is divided into two parallel signal processing parts: the micro-Doppler-based motion feature extraction part and the Doppler frequency-based vital sign feature extraction part.

The I and Q signals reflected from the object are sampled through an ADC (Analog Digital Converter). In this paper, we set the ADC sampling rate to 1 kHz, which is a high enough value to digitalize the received signal form the CW radar system.

The corresponding digitized data are inputted into two signal processing parts. In the micro-Doppler-based motion feature extraction part, we can extract two feature vectors (*x1* and *x2*) that can determine whether or not human motion appears. Details are presented below in Figures 4 and 5.

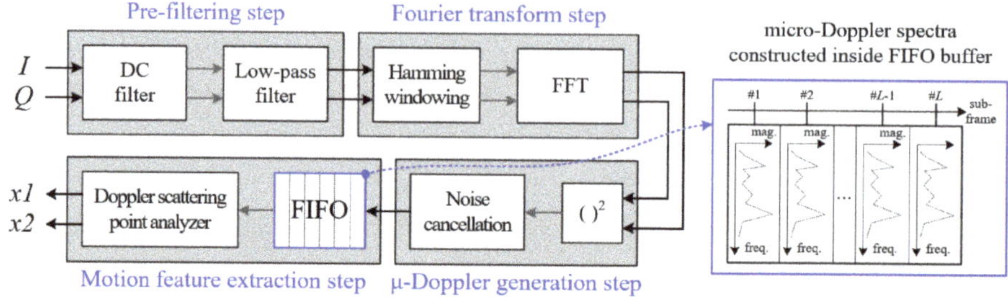

Figure 4. Detail algorithm steps for micro-Doppler-based motion feature extraction.

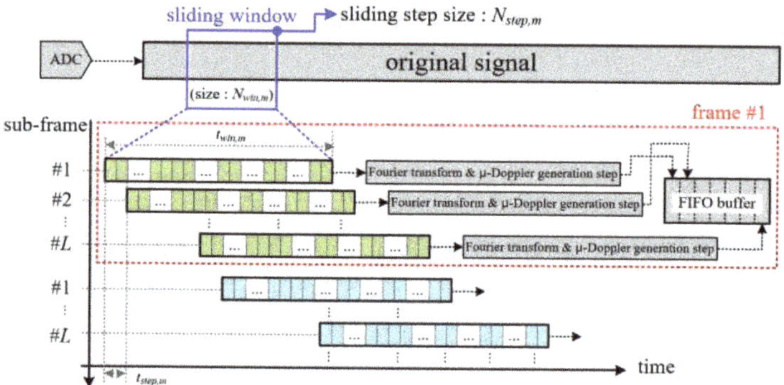

Figure 5. Concept of sliding window-based micro-Doppler generation for motion feature extraction.

In the Doppler frequency-based vital sign feature extraction part, the third vector *x3* is extracted to determine whether the vital signs exist. Figure 6 shows the detailed flow of the signal processing together with a corresponding description.

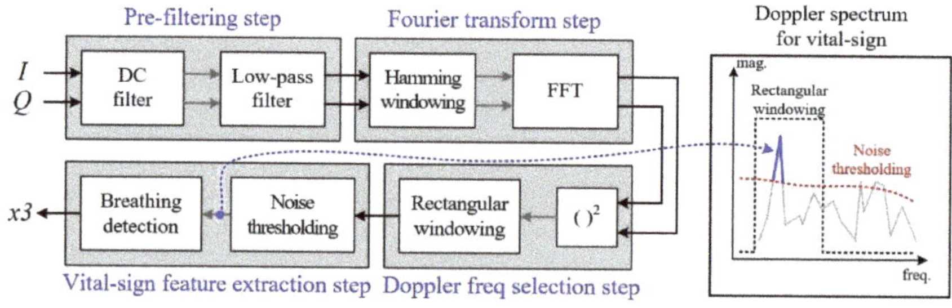

Figure 6. Detailed algorithm steps for Doppler frequency-based vital sign feature extraction.

Figure 4 shows the proposed signal processing flow of the micro-Doppler-based motion feature extraction part. The process has four steps, as shown below.

First, in the pre-filtering step, from the I and Q signals, the DC and high-frequency components are removed using a DC filter and an LPF (Low Pass Filter), respectively. In the paper, the DC filter is implemented by subtracting the average value from the raw signal. We also design the LPF on the 64th order with a cut-off frequency $f_{h,m}$ of 30 Hz.

Second, we conduct the Fourier transform by multiplying the coefficients of a Hamming window with $N_{win,m}$ to suppress the side-lobe of the frequency spectrum and the FFT (Fast Fourier Transform) with K_m points for conversion to the frequency domain.

For a detailed explanation of the Fourier transform step, we illustrate the data processing timing flow of the original raw signal in Figure 5.

We employ the STFT (Short Time Fourier Transform) technique based on a sliding window. Because the required Doppler frequency resolution used to separate motion and vital signs is set to approximately 0.5 Hz, we select a window size $N_{win,m}$ of 2000, which means that the measurement time has a $t_{win,m}$ value of 2 s. Moreover, because the maximum Doppler frequency of a passenger's motion is about 10 Hz, we set the sliding step time $t_{step,m}$ to 0.1 s for sliding window, indicating $N_{step,m}$ sample of 100.

Thus, in this paper, a Doppler spectrum with 2048 FFT points is generated from 2000 samples of the original signal. This procedure is repeated continuously with a sliding window in 0.1 s steps, as shown in Figure 5.

Next, in the micro-Doppler generation step, the magnitude of the Doppler frequency spectrum is calculated using the root-square function, and then the background noise is removed using the previously measured noise spectral distribution information. The generated Doppler frequency spectrum is saved into FIFO (First Input First Output) memory, and the micro-Doppler image is composed by each Doppler spectrum such as the right side of Figure 4. Here, FIFO memory can store as many Doppler spectra as the number of sub-frames L.

Finally, in the motion feature extraction step, as shown in Figure 4, we can obtain two feature vectors by analyzing the distribution of the Doppler scattering points over all sub-frames. This procedure is carried out in a Doppler scattering point analyzer, which will be described later and is shown in Section 2.3.

Figure 6 shows the procedure of the Doppler frequency-based vital sign feature extraction part, which is a simple technique. The process has also four steps, as shown below.

The roles of the pre-filtering step and the Fourier transform step are identical to those in the case of Figure 4. However, the LPF is on the 32th order with a cut-off frequency $f_{h,v}$ of 1 Hz. Moreover, the length of the Hamming window and the number of FFT points are expressed by $N_{win,v}$ and K_v, respectively.

We also employ STFT technique with a sliding window, as shown in Figure 7. Here, because the maximum Doppler frequency of human vital signs is less 1 Hz, the sliding step time $t_{step,v}$ is set to 1 sec, with reference to a $N_{step,v}$ sample of 1000.

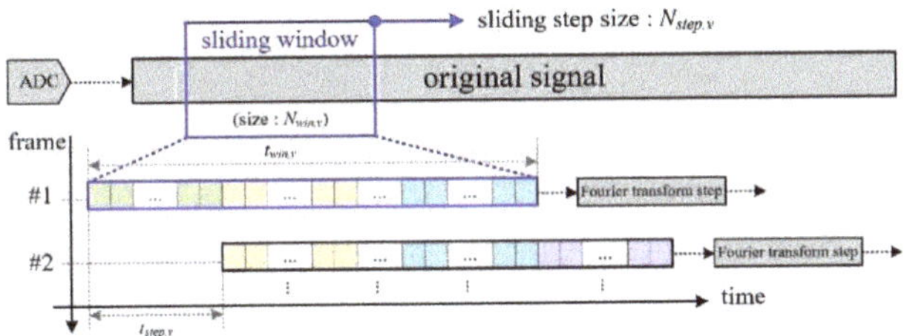

Figure 7. Concept of sliding window-based Doppler frequency extraction for vital sign feature extraction.

Moreover, for vital sign signals, because the minimum frequency is bounded in the range of 0.1 Hz to 0.5 Hz, the window size $N_{win,v}$ is set to 8000. Thus, we measure the received signal during a $t_{win,v}$ time of 8 s in order to confirm the presence of a breathing signal. As a result, the FFT point is set to 8192.

Next, in the Doppler frequency selection step, the absolute values are calculated as the Doppler spectrum, and rectangular windowing is applied in the frequency domain in order to consider only values below the $f_{h,v}$ frequency, such as the right side of Figure 6.

Finally, in vital sign feature extraction step, as shown in Figure 6, we can set a threshold of scattering points with a magnitude greater than the noise, and we determine vital signs based on survival values. The details pertaining to this are described later and are shown in Figure 8.

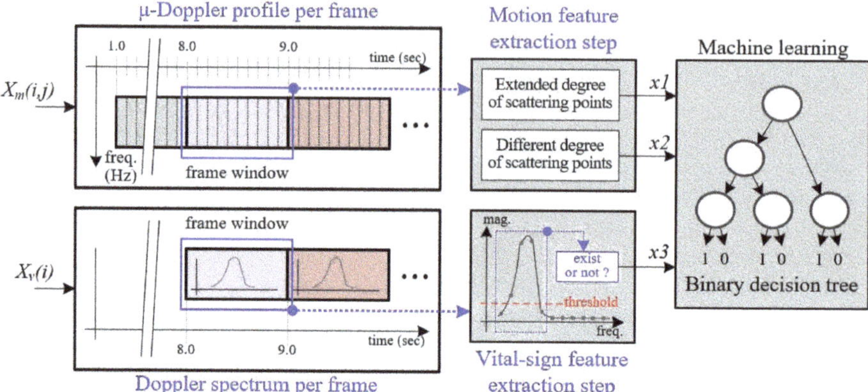

Figure 8. Process of extracting both motion and vital sign features from the micro-Doppler and Doppler spectra with the same time interval.

As explained in Figures 5 and 7, while the Doppler spectra for motion and vital signs are generated every 0.1 s and 1 s, respectively, according to the step time of the sliding window. That is, while one Doppler spectrum for vital sign detection is generated, ten Doppler spectra for motion detection are completed. Thus, in order to synchronize the two parts and effectively recognize human motion, we generate a micro-Doppler image using ten sub-frame spectra.

2.3. Proposed Feature Vector Extraction Scheme

As explained above, because the maximum Doppler frequency of human vital signs is less 1 Hz, in our design, the update time of vital sign detection is 1 sec. In a previous study [13], vital signs were analyzed every one second. Thus, in this paper, we generate a micro-Doppler image for human motion detection every second to synchronize the update time with the vital sign extraction point.

Figure 8 shows the timing diagram used to synchronize the micro-Doppler image for motion detection and the Doppler spectrum for vital sign detection. In this case, the data collection time is initially set to 8 s to generate the first frame's Doppler spectrum for vital sign detection.

For a detailed explanation of the motion feature extraction process, we define the data stored in FIFO as $\{X_m(i, j), i = 1 \sim K_m, j = 1 \sim L\}$, where K_m is the Doppler-bin size and L is the number of sub-frames. In this paper, because we set K_m to 2048 and L to 10, the size of the micro-Doppler image in one frame is 2048 by 10.

To extract the first and second feature vectors, we first count the number of scattering points stored in FIFO memory every sub-frame, which is expressed as Equation (1). From $Y(j)$, we can obtain the corresponding count value in the jth sub-frame.

$$Y(j) = \text{Count}\{X_m(i, j) > 0\} \text{ for } i = 1 \sim K_m \tag{1}$$

While the vital sign signals of a non-moving passenger or the Doppler signal of another moving object have a narrow distribution of the frequency spectrum, a wide Doppler spectrum can appear when a human moves on the seat of the vehicle. In this paper, to express these characteristics, we define the 'extended degree of scattering points' as the first feature, x_1,. This is expressed by Equation (2), which indicates the average number of scattering points of all sub-frames.

In addition, while the Doppler signals from the breathing of a still human are mostly maintained over time, the passenger's movements may not be continuous. Thus, the distribution of the Doppler scattering points can vary over the sub-frames. Thus, in this paper, we define these characteristics as the 'different degree of scattering points', i.e., as the second feature x_2. Equation (3) is used to solve x_2, which is obtained by averaging the differences between the numbers of scattering points of two consecutive sub-frames.

These two feature vectors make it possible to determine whether or not a moving object on a seat in a vehicle is human.

$$x_1 = \left\{ \sum\nolimits_{j=1}^{L} Y(j) \right\} / L \tag{2}$$

$$x_2 = \left\{ \sum\nolimits_{j=1}^{L-1} |Y(j+1) - Y(j)| \right\} / (L-1) \tag{3}$$

Next, we define the Doppler frequency spectrum as $\{ X_v(i), i = 1 \sim K_v \}$ to process the results of the Doppler frequency-based vital sign feature extraction, where K_v is the Doppler-bin size. That is, we count the number of scattering points with values greater than reference value for noise thresholding. The reference value can be obtained as the measured noise signal in blank space.

$$Z = \text{Count}\{X_v(i) > \text{reference value}\} \text{ for } i = 1 \sim K_v \tag{4}$$

Using the results of Equation (4), we define x_3 as the 'presence of vital signs', and the third feature is expressed in Equation (5). When using the feature vector x_3, we can determine whether or not a non-moving or slightly moving object is a human.

$$x_3 = \begin{cases} \text{if } Z > 0, \text{ then logic '1'} \\ \text{else logic '0'} \end{cases} \tag{5}$$

The three extracted feature vectors are fed into the machine learning engine for learning and testing. In this paper, we employ a BDT (Binary Decision Tree) as the machine learning method.

The BDT is a popular and simple machine learning algorithm based on a sequential decision process because a feature is evaluated as one of two branches, which is selected starting from the root of the tree. Thus, we can easily implement the BDT in an embedded system for machine learning based on the three features proposed in this paper, using only the "if–else" syntax in real time [15].

3. Measurement Results

3.1. Radar Sensor and Measurement Environment

To verify the proposed algorithm, we established a test-bed in the DGIST lab, as shown in Figure 9. The test-bed is composed of a Doppler radar FEM (Front-end Module) with antennas, a DAQ (Data Acquisition) module, a power supply, and a PC.

The radar antennas are mounted to face the vehicle seats, and the transmitted and received ports are connected to the FEM using SMA (Sub Miniature type-A) cables. The received baseband signals are logged by the DAQ module, the data are sent to the PC through a USB (Universal Serial Bus) interface. To control DAQ module and obtain the data in real time, we developed DAQ software using NI's LabVIEW tool on the PC.

For the purposes of this paper, the FEM and antennas were manufactured by Yeungnam University. In the FEM, a VCO (Voltage Controlled Oscillator) is added, which is different from the board in an earlier circuit version [12]. A photo is shown in Figure 10.

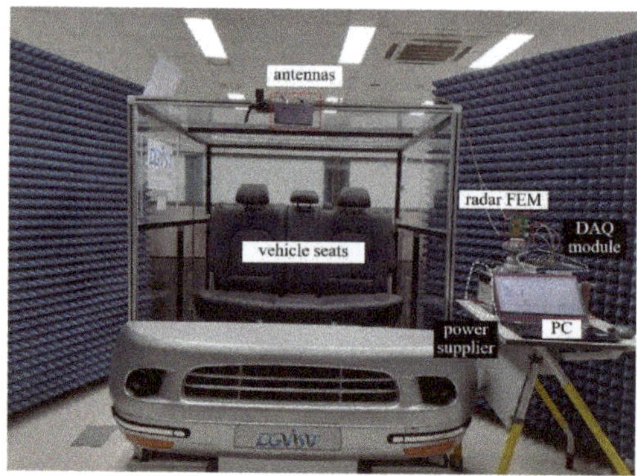

Figure 9. Photo of the test-bed built in the DGIST lab.

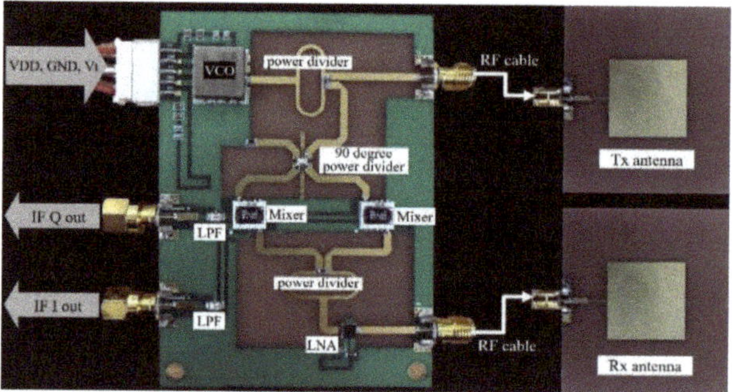

Figure 10. Photo of the 2.45 GHz CW radar prototype.

The detailed specifications are presented in Table 1. The center frequency is 2.45 GHz and the FOV (Field of View) of the antenna is 80 degrees. In NI's DAQ module used here, we set the sampling rate to 1 kHz and the input dynamic range to −5~5 V through the LabVIEW tool.

Table 1. Parameters of the radar system used in this paper.

Parts		Specifications	Units	Symbols	Values
FEM/antennas		Center frequency	GHz	f_c	2.45
		Field of view	Degree	-	80
DAQ module		ADC frequency	MHz	f_s	1
		Dynamic range	V	-	−5~5
Signal processing	Motion detection	Cut-off frequency	Hz	$f_{h,m}$	10
		FFT point	point	K_m	2048
	Vital sign detection	Cut-off frequency	Hz	$f_{h,v}$	1
		FFT point	point	K_v	8192

In addition, the LPF and FFT parameters for the signal processing described above are also shown in Table 1. For the 2.4 GHz Doppler radar system, because the Doppler frequencies of passenger motion and breathing do not exceed 10 Hz and 1 Hz, we select the cut-off frequencies shown in Table 1. The FFT points are also selected such that they support the resolution of the Doppler frequency.

3.2. Measurement Scenarios

To verify the proposed human recognition scheme, we designed eight cases as scenarios to be carried out on the test-bed. A photo of each case is presented in Figure 11. In this paper, we measure the radar signal for 60 s in every case. That is, in each case, measurements were conducted for 60 frames. However, in each case, the number of the extracted feature vectors is 52, because of the initial data collection time of 8 s, as described in Section 2.3.

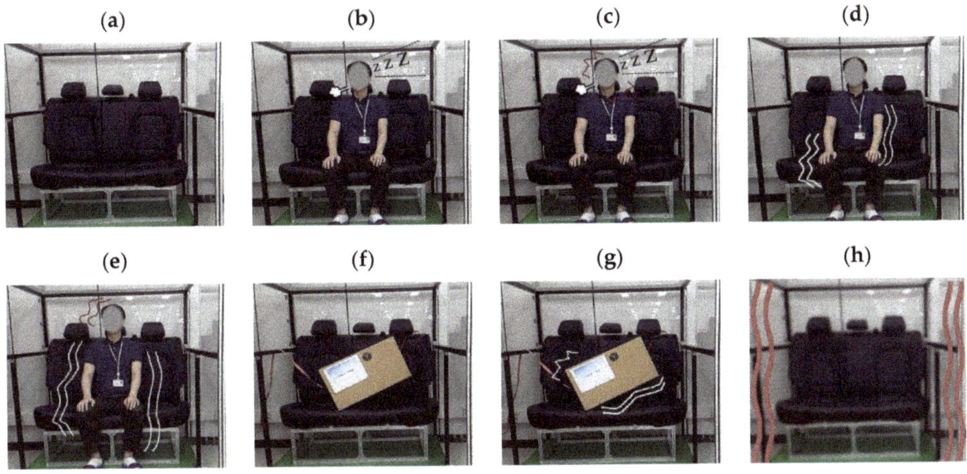

Figure 11. (**a–h**) Test scenarios represented as eight cases in a vehicle.

In this paper, we only consider humans as living creatures. That is, we do not discuss other living creatures such as companion animals, because the recognition of the passenger as a human is most important. Moreover, we considered the seats in the vehicle, a box on a seat, and the vehicle body itself as inanimate objects. A detailed description of the scenario is given below.

- Case #1: Empty seat without any objects.
- Case #2: A still human on a seat; for example, a passenger who is sleeping without any motion on the seat.
- Case #3: A human moving his neck slightly on a seat; for example, a passenger who is dozing with neck movement.
- Case #4: A human making slight motion on a seat, such as a passenger who is looking around, talking with hand gestures, or has light body movements.
- Case #5: Humans making more than slight motions on the car seat, such as a passenger who is listening to music with the head or body moving slightly in a wavy motion, shaking leg, keeps moving his body, or has additional body movements.
- Case #6: Still box on a seat.
- Case #7: Slightly wobbly box on a seat, such as when a box on the seat is lightly shaken by vehicle vibration; this scenario is virtually simulated by connecting a string to the box.
- Case #8: Vibrating vehicle such as when the vehicle itself vibrates while driving; this scenario is virtually simulated by actually shaking the test bed.

For the measurements of cases #2–#5, we conducted the experiments with three males. The characteristics of these participants are given below. The photos in Figure 11 show human #2. In the experimental results in Sections 3.3 and 3.4, we present the results from human #2.

- Human #1 is 27 years old, 183 cm tall, and weighs 78 kg.

- Human #2 is 37 years old, 176 cm tall, and weighs 80 kg.
- Human #3 is 47 years old, 179 cm tall, and weighs 90 kg.

The CW Doppler radar operates at the 2.45 GHz ISM band, which is freely used for the purpose of industrial, science, and medical applications. The transmitted power of the radar is less than 5 dBm, which meets the regulation for human bodies. All the subjects joining in the experiment agreed with protocol, procedure, and process in the measurement.

3.3. Pre-Processing Results

Figure 12 shows the received raw signal for a set time of eight seconds for all cases, where the x-axis and y-axis correspondingly indicate the time (s) and amplitude (voltage level).

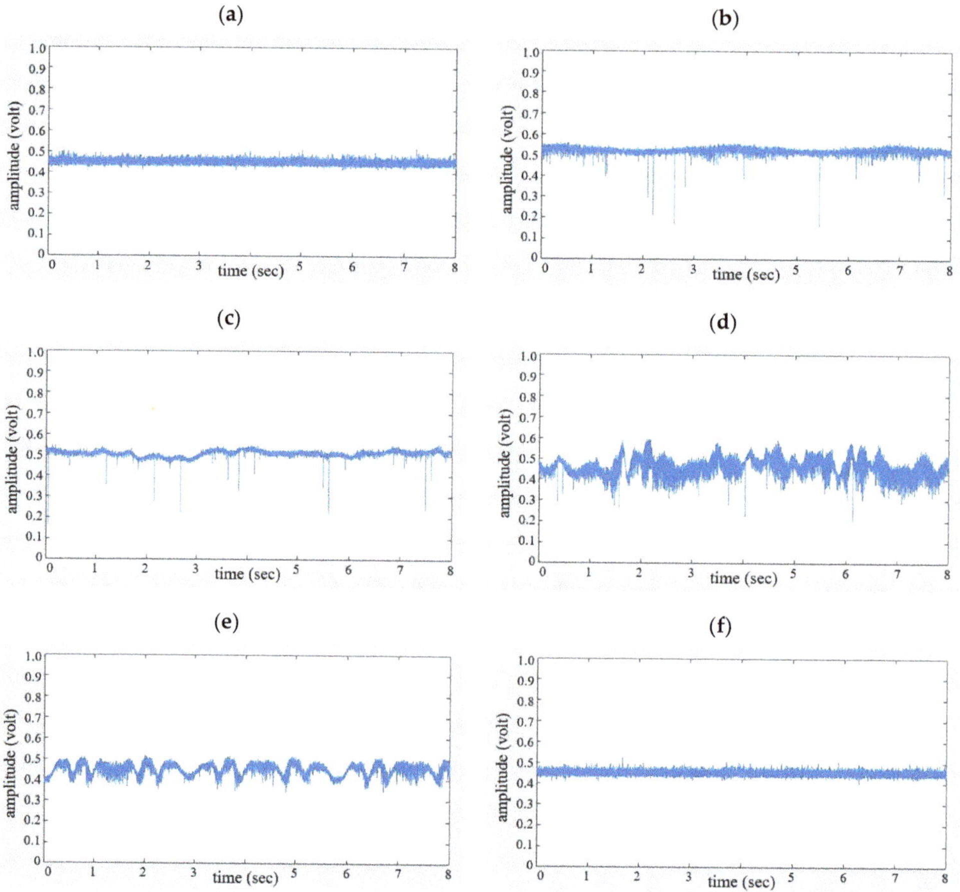

Figure 12. *Cont.*

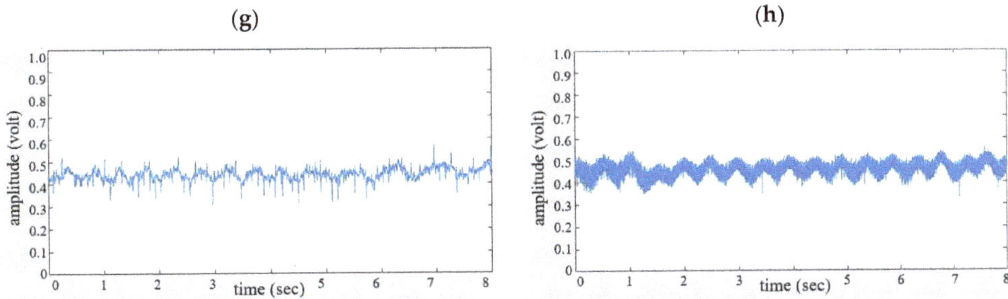

Figure 12. Received signals in the time domain; (**a**–**h**) indicate cases #1~#8.

In Figure 11a,f, only noise is found, since there are no moving objects (cases #1 and #6). While a breathing signal appears as a sine wave for the human without motion in Figure 11b, we find that the breathing and motion signals are mixed due to the slight movement of the neck (case #3) in Figure 11c.

In Figure 11d,e, because more human motion exists, aperiodic signals are displayed with a higher frequency than the respiration of a human.

Finally, Figure 11g,h show periodic vibration signals that appear due to the shaking of the box and vehicle (cases #7 and #8).

Figure 13 shows a micro-Doppler image generated from the micro-Doppler-based motion feature extraction part described in Figure 4. Figure 12a–h are the signal processing results of Figure 12, showing the results for the eight aforementioned cases. Here, the x-axis is the time (s) and the y-axis is the frequency (Hz).

In Figure 12a,f, no scattering points higher than the background noise are seen because there is no moving object.

In Figure 12b,c, because the human has no movement or only slight motion, mostly breathing signals are extracted. In this case, Doppler spectra with a narrow shape appear almost continuously over time.

Interestingly, in Figure 12g,h, sharp patterns appear, similar to those in Figure 12b,c. This occurs because other objects do not have multiple scattering points.

On the other hand, in Figure 12d,e, a wide distribution of scattering points is found due to various components of human movement. In addition, it can be seen that the distribution of the scattering points varies over time.

Based on these micro-Doppler images, we can extract two feature vectors x_1 and x_2 as the 'different degree of scattering points' and the 'extended degree of scattering points' through processing, as shown in Figure 8.

Figure 14 shows the Doppler spectra of the Doppler frequency-based vital sign feature extraction part shown in Figure 6. These results are also measured based on the signals in Figure 12 for all eight scenarios. Here, the x-axis is frequency (Hz) and the y-axis denotes the magnitude. In addition, the red-dotted boxes indicate the area of the frequency below 1 Hz, occupied by the breathing signal.

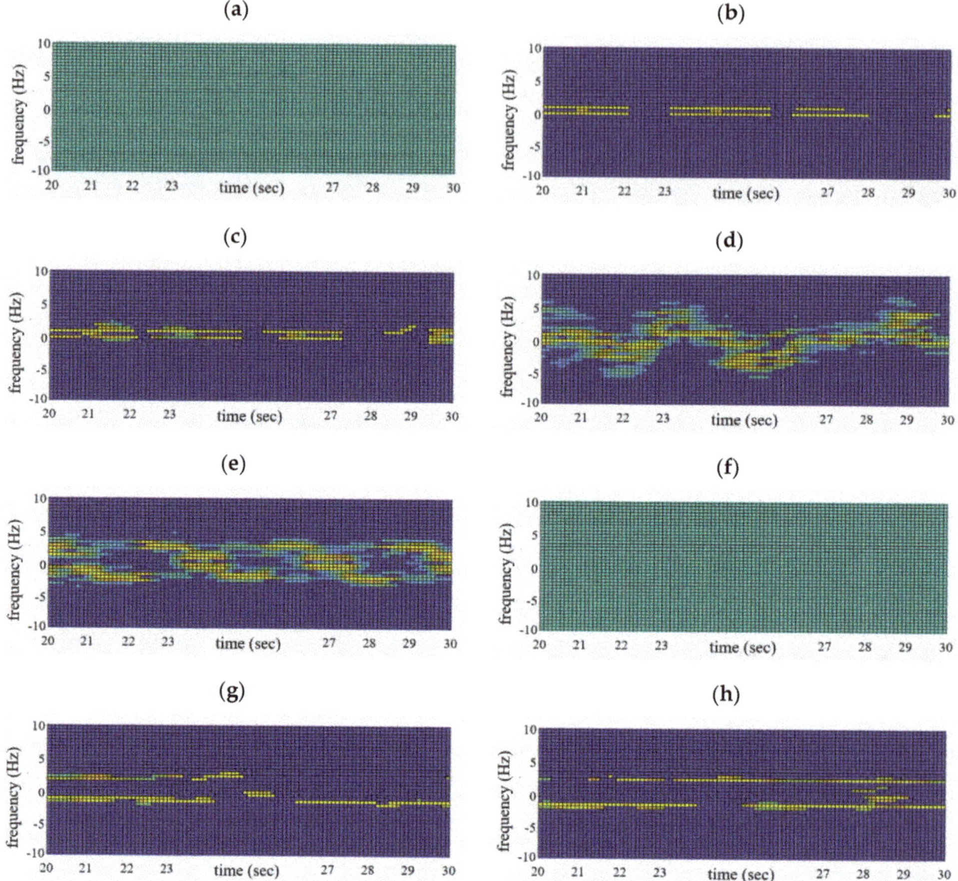

Figure 13. Extracted micro-Doppler image of μ-Doppler-based motion feature extraction part; (**a**–**h**) correspondingly indicate cases #1 to #8.

In Figure 14a,f, no dominant spectrum is found due to the absence of motion or vital sign. In Figure 14b,c, the Doppler spectrum is sharp in the red box due to the breathing signal. However, the spectrum of Figure 14c is slightly widened compared to that in Figure 14b due to the motion of the neck.

In Figure 14g,h, sharp-type Doppler spectra are also found, but most of them are located outside of 1 Hz.

On the other hand, in Figure 14d,e, we find that the Doppler spectra are spread across the red box. Occasionally, maximum peaks can be found at less than 1 Hz according to the Doppler volume of human movement.

Based on these Doppler spectra, we can determine whether or not vital signals exist, and obtain the third feature vector x_3. In this paper, a simple algorithm that can extract the breathing signal is employed. If we use a very fine algorithm, we can extract vital signs with a high detection probability.

Figure 15 shows two motion features extracted from micro-Doppler images of Figure 13 for eight cases. Figure 15a,b indicate the first feature x_1 as the 'extended degree of scattering points' and x_2 as the "different degree of scattering point". Here, the x-axis is the time (sec) and the y-axis is the number of the detected scatters. The results for cases #1 to #8 are correspondingly represented by the green-solid, red-solid, red-dotted, blue-solid, blue-dotted, green-dotted, black-solid, and black-dotted lines.

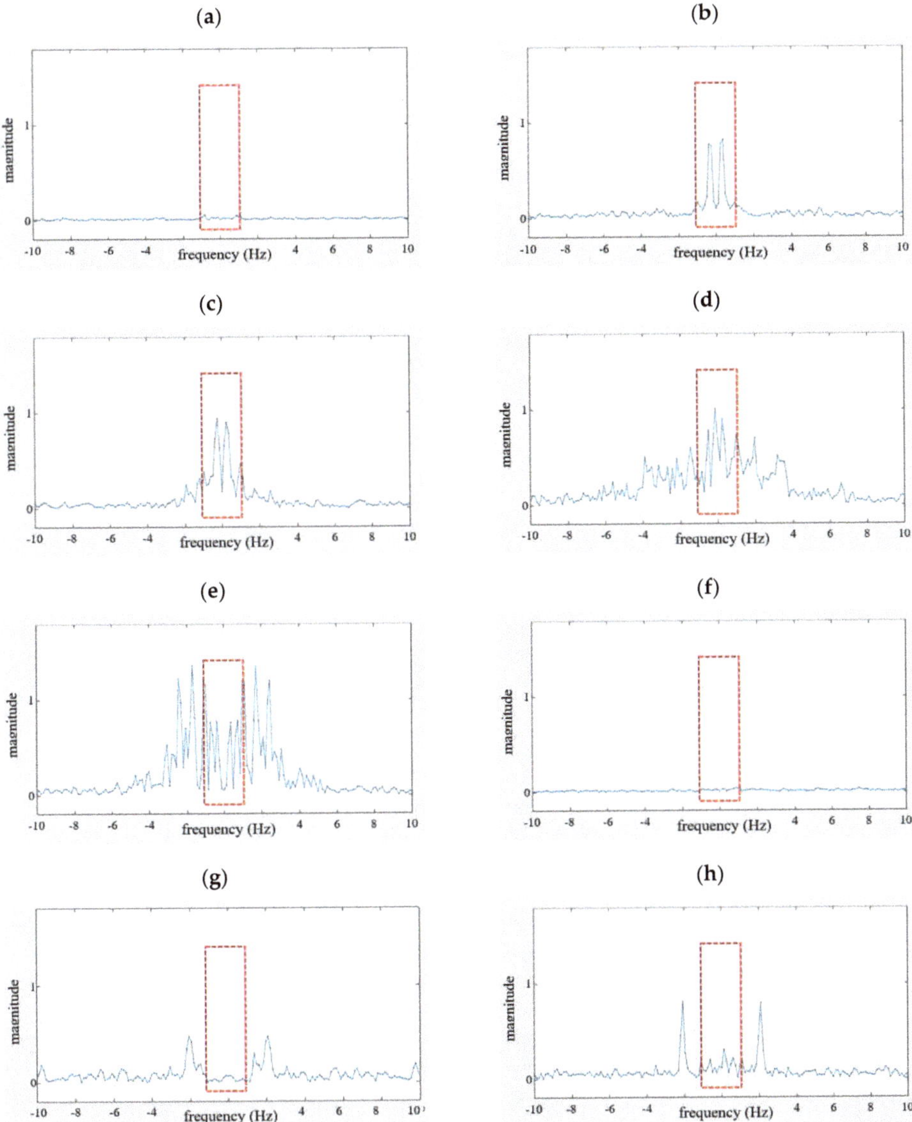

Figure 14. Doppler spectrum extracted from the Doppler frequency-based vital sign feature extraction part; (**a–h**) indicate cases #1 to #8.

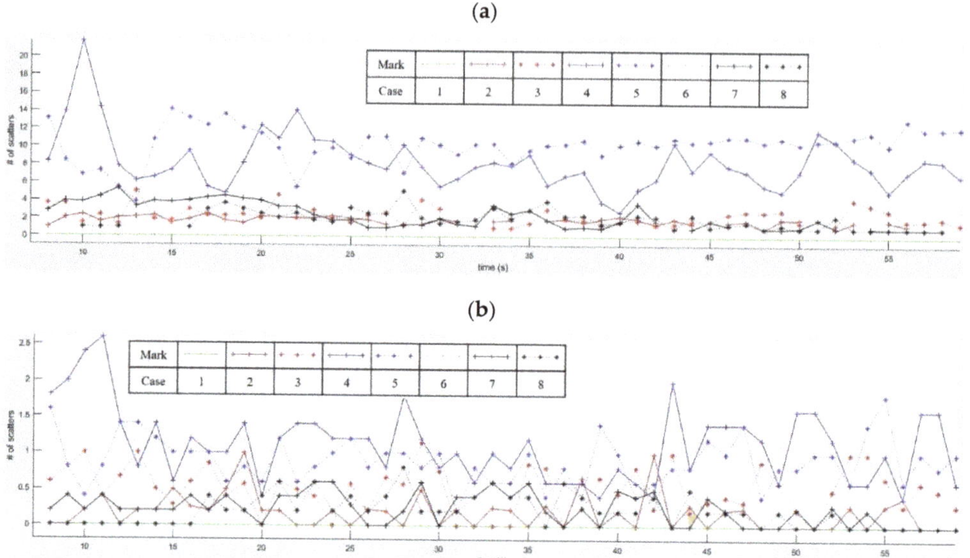

Figure 15. Motion feature extraction results for eight cases: (**a**) results of the extended degree of scattering points and (**b**) the results of the different degree of scattering point.

Figure 15a shows that the number of scatters in cases #2 and #3 is mostly lower compared to cases #4 and #5 due to the different levels of human movement. That is, for a still or slightly moving human, only a few scattering points are reflected.

In cases #7 and #8, we can find similar patterns to those in cases #2 and #3. In cases of inanimate objects, the Doppler spectra are narrow, because multiple components do not exist.

Finally, when there is no motion component, the number of extracted features is 0, such as in cases #1 and #6.

In Figure 15b, when the passenger is moving on a seat for cases #4 and #5, we find that the distribution of the Doppler spectrum varies more than those of a still human (case #2), a slightly moving human (case #3), and other objects (cases #7 and #8).

Figure 16 shows the extracted Doppler frequency spectra for the third vital sign feature in Figure 14. That is, Figure 16a,h indicate the presence or absence of vital sign. For all cases, the x-axis and the y-axis denote the time (s) and logic value (true or false), respectively.

As shown in Figure 16a,f, no signal is detected in the cases without motion

In Figure 16b,c, vital signals are recognized at all times, despite the fact that the passenger moves his neck only slightly. However, from the results in Figure 16d,e, we find that breathing signs may or may not be extracted depending on the movement level of the human.

Finally, for inanimate objects, we can find that vital signals are mostly not detected, as shown in Figure 16g,h. However, in the results, false detections occasionally occur due to noise. This problem will be resolved by employing a fine breathing detection algorithm in the future.

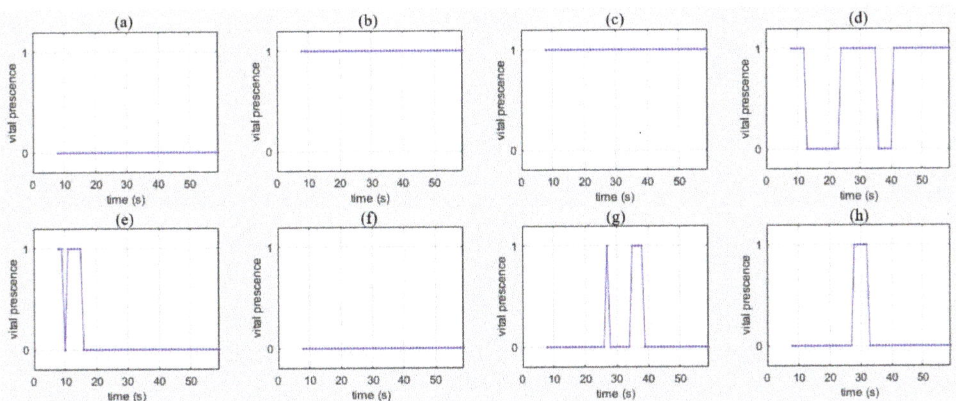

Figure 16. Vital sign feature extraction results; (**a–h**) indicate cases #1 to #8.

3.4. Proposed Feature-Based Human Recognition Results

Figure 17 presents the three-dimensional distributions of the features extracted from the micro-Doppler image and the vital sign frequency spectrum.

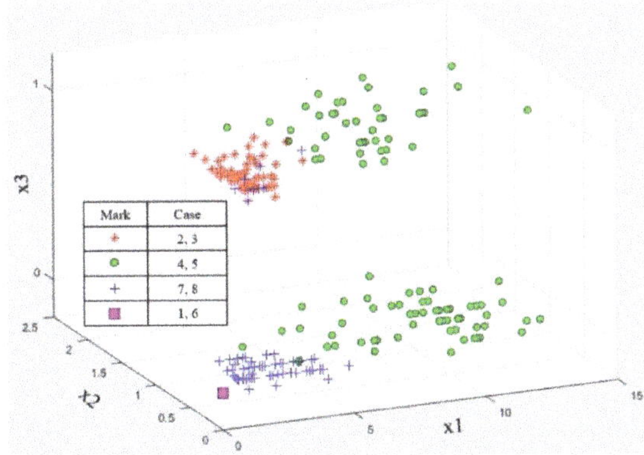

Figure 17. Three-dimensional distribution of three features extracted from the proposed algorithm scheme for a human and other objects.

In cases #1 and #6, without any moving object, the three features are positioned at zero, as shown by the purple boxes.

We can distinguish between a human with no or little motion (cases #2 or #3) and an inanimate object with movement (cases #7 and #8) using only the vital sign feature x_3. Here, cases #2 and #3 are displayed with red-star marks, and the blue-cross marks are used to present cases #7 and #8. However, as mentioned above, even in cases #7 and #8, a few incorrect results appear, as if vital signals are detected due to noise.

As shown by the green circle marks in cases #4 and #5, because the x_3 values for the passenger with movement are distributed between 1 and 0, it is impossible to determine whether the detected object is human or not if using only x_3. However, in these two cases, the first and second features extracted from the micro-Doppler image are positioned in an area far from the origin, while the results for the inanimate object appear around the origin. Thus, when using x_1 and x_2, we can distinguish between a human with motion and other objects with movement.

In this paper, we use three feature vectors to train and test the process using machine learning with the BDT, as shown in Figure 18. The procedure we used for all programming for machine learning and verification is described below. Here, we coded all procedures using the Matlab library.

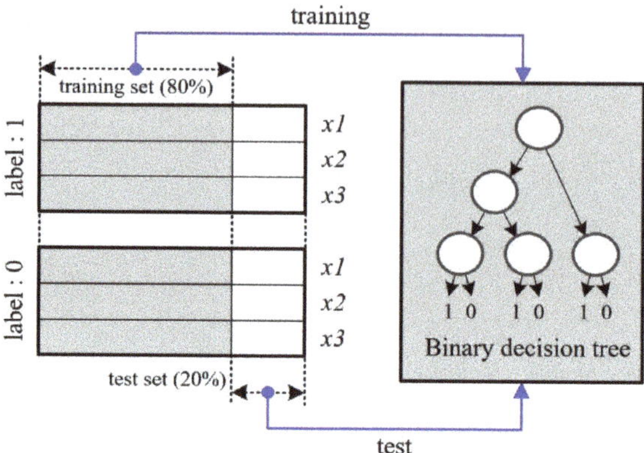

Figure 18. Structure of features comprising the training data set and the test set for machine learning.

- We labeled the features of the actual human as '1' for 208 frames and other cases as '0' for 208 frames, respectively. Here, one frame was measured every one second, as shown in Figure 8.
- We then randomly separated the three features for the human cases and the others, with 80% for the training data set and 20% for the test set. This means that 80% of the total 208 frames with label '1' were used for the training set, with the remaining 20% being allocated to the test set. Moreover, among the feature vectors of 208 frame times with the label '0', the data of 166 s and 42 s were used for training and test, respectively.
- We optimized the machine learning engine of the BDT library via a 30-trial loop with the training data set.
- We input the test set into the optimized machine learning engine.
- We repeated the three-step procedure described above ten times, while also dividing the data set, training, and testing steps.
- We checked the performance by averaging the results of the ten aforementioned trials.

In the typical methods [7,12], vital sign monitoring of breathing in vehicle applications is used. That is, the sampled radar echo signal is analyzed for the presence of periodic breathing while separating vital signs form background noise. Thus, previous works considered only the scenario of a human being asleep in a vehicle. In this paper, we define instances that use only vital sign signals as the typical method.

In Table 1, the human recognition performances outcomes are presented for the typical method and with the proposed algorithm. In the typical algorithm, only the vital signals are used in cases for a human with no motion and with slight motion. However, in the proposed algorithm, we use not only the characteristics of the Doppler scattering points, but also the breathing signals.

We present two performance metrics: the classification accuracy (%) and the classification error rate (%).

The classification accuracy indicates whether or not an actual human has been accurately classified as a human. On the other hand, the classification error rate represents the rate at which an inanimate object is mistakenly determined to be a human.

While the classification accuracy of the typical algorithm using only vital signs is approximately 70%, the classification accuracy of the proposed method is improved to 98.6%, as shown in Table 2. That is, the performance is enhanced by nearly 28%.

Table 2. Classification performance for human recognition.

Metrics Algorithm	Performance	Classification Accuracy (%)	Classification Error Rate (%)
typical method		70.7	4.8
proposed method		98.6	5.3

Regarding the classification error rate, the performance of the proposed method is decreased by 0.5% compared to the typical method. In this paper, because we employ a very simple algorithm to detect vital signs, the noise of an inanimate object can be occasionally incorrectly recognized as vital signs. If the vital sign detection algorithm is advanced in the future, this problem will be resolved.

In this paper, we conducted this experiment with three people, and similar performance outcomes were obtained. This occurred because the measuring distance is very close, at about 1 m, and the shapes of the human bodies of the participants are similar.

4. Conclusions

In this paper, we defined three new features and proposed a human recognition scheme based on machine learning using a CW radar sensor. To do this, we initially measured the 'extended degree of scattering points' from micro-Doppler images, after which we calculated the mean of the Doppler reflection points over sub-frames. Second, in order to extract the 'different degree of scattering points', we calculated the mean of the difference in the Doppler reflection points between two successive sub-frames.

While the two feature vectors described above were meant to recognize a human's motion, the last feature vector is for human vital sign recognition. Hence, we defined the 'presence of vital signs' as extracted from the Doppler frequency spectrum as the breathing signal of a human.

To verify the performance of the proposed algorithm, we built a test-bed similar to the interior of an actual vehicle and defined eight cases, consisting of non-moving objects, a still or moving human, and inanimate moving object scenarios. For these cases, we used a commercial real-time DAQ module and a 2.45 GHz CW radar front-end module with antennas developed by Yeungnam University. Then, in order to extract three features from the received radar signals, we obtained the raw data using the test-bed.

The extracted features were used as input data for a BDT as machine learning engine, and we verified the proposed algorithm through randomly repeated verification trials.

The results with the typical method using only vital signs show that the classification accuracies for a human were 70.7%. However, with the proposed human recognition scheme using motion and vital sign features, the classification accuracy was found to be 98.6%. That is, compared to the typical method, the performance of the proposed method is improved by approximately about 28%.

Moreover, because the proposed algorithm has very low complexity, we can implement a passenger detection radar system with a simple structure.

In the future, using radar sensors with multiple receiving antennas, we will conduct research to determine the presence and status of occupants for each seat. In the addition, we plan to employ a new vital sign detection algorithm to improve the classification error rate for humans. We will also verify the proposed algorithm together with the various types of human forms. Moreover, we will install the test-bed in an actual vehicle in order to verify the proposed recognition scheme more practically.

Author Contributions: E.H. handled the design and development of the proposed algorithm, contributed to the creation of the main ideas in the paper, and handled the writing of the manuscript. Y.-S.J. focused on the experimental task (i.e., setting up the system, the design and realization of the experiments). J.-R.Y. designed the radar front-end module and antenna, and J.-H.P. fabricated and tested the hardware of the radar sensor. All authors conducted data analysis and interpretation tasks together. All authors have read and agreed to the published version of the manuscript.

Funding: This research received no external funding.

Acknowledgments: This research was supported by the DGIST R&D Program funded by the Ministry of Science and ICT, Korea. (No: 20-IT-02, Title: Development of core technologies for environmental sensing and software platform of future automotive)

Conflicts of Interest: The authors declare no conflict of interest.

References

1. Available online: https://www.euroncap.com/ (accessed on 30 October 2020).
2. Automotive News Europe. Child-Detection Safety Technology May Get Mandate. Available online: https://europe.autonews.com/automakers/child-detection-safety-technology-may-get-mandate (accessed on 30 October 2020).
3. Ismail, N.H.F.; Husain, N.A.; Mansor, M.S.F.; Baharuddin, M.M.; Zaki, N.M.; Husain, M.A.; Ma'aram, A.; Wiyono, A.S.; Chaiyakul, T.; Ahmad, Y. Child Presence Detection System and Technologies. *J. Soc. Automot. Eng. Malays.* **2019**, *3*, 290–297.
4. Chongpyo, C.; Gangchul, K.; Youngdug, P.; Wookhyun, L. The development of an energy-efficient heating system for electric vehicles. In Proceedings of the 2016 IEEE Transportation Electrification Conference and Expo, Asia—Pacific (ITEC), Busan, Korea, 1–4 June 2016; pp. 883–885.
5. Elbanhawi, M.; Simic, M.; Jazar, R. In the passenger seat: Investigating ride comfort measures in autonomous cars. *IEEE Intell. Transp. Syst. Mag.* **2015**, *7*, 4–17. [CrossRef]
6. Hyun, E.; Jin, Y.; Lee, J. A Pedestrian Detection Scheme Using a Coherent Phase Difference Method Based on 2D Range-Doppler FMCW Radar. *Sensors* **2016**, *16*, 124. [CrossRef]
7. Da Cruz, S.D.; Beise, H.-P.; Schroder, U.; Karahasanovic, U. A Theoretical Investigation of the Detection of Vital Signs in Presence of Car Vibrations and RADAR-Based Passenger Classification. *IEEE Trans. Veh. Technol.* **2019**, *68*, 3374–3385. [CrossRef]
8. Yang, Z.; Bocca, M.; Jain, V.; Mohapatra, P. Contactless Breathing Rate Monitoring in Vehicle Using UWB Radar. In Proceedings of the 7th International Workshop on Real-World Embedded Wireless Systems and Networks, Shenzhen, China, 4 November 2018; pp. 13–18.
9. Lim, S.; Lee, S.; Jung, J.; Kim, S.-C. Detection and Localization of People inside Vehicle Using Impulse Radio Ultra-Wideband Radar Sensor. *IEEE Sens. J.* **2020**, *20*, 3892–3901. [CrossRef]
10. Kim, B.; Jin, Y.; Lee, J.; Kim, S. Low-Complexity MUSIC-Based Direction-of-Arrival Detection Algorithm for Frequency-Modulated Continuous-Wave Vital Radar. *Sensors* **2020**, *20*, 4295. [CrossRef] [PubMed]
11. Wang, Y.; Wang, W.; Zhou, M.; Ren, A.; Tian, Z. Remote Monitoring of Human Vital Signs Based on 77-GHz mm-Wave FMCW Radar. *Sensors* **2020**, *20*, 2999. [CrossRef] [PubMed]
12. Diewald, A.R.; Landwehr, J.; Tatarinov, D.; Cola, P.D.M.; Watgen, C.; Mica, C.; Lu-Dac, M.; Larsen, P.; Gomez, O.; Goniva, T. RF-based child occupation detection in the vehicle interior. In Proceedings of the 2016 17th International Radar Symposium (IRS), Krakow, Poland, 10–12 May 2016.
13. Kim, J.-Y.; Park, J.-H.; Jang, S.-Y.; Yang, J.-R. Peak Detection Algorithm for Vital Sign Detection Using Doppler Radar Sensors. *Sensors* **2019**, *19*, 1575. [CrossRef]
14. Hyun, E.; Jin, Y. Human-vehicle classification scheme using doppler spectrum distribution based on 2D range-doppler FMCW radar. *J. Intell. Fuzzy Syst.* **2018**, *35*, 6035–6045. [CrossRef]
15. Hyun, E.; Jin, Y. Doppler-Spectrum Feature-Based Human–Vehicle Classification Scheme Using Machine Learning for an FMCW Radar Sensor. *Sensors* **2020**, *20*, 2001. [CrossRef] [PubMed]

Publisher's Note: MDPI stays neutral with regard to jurisdictional claims in published maps and institutional affiliations.

Letter

Effect of Filtered Back-Projection Filters to Low-Contrast Object Imaging in Ultra-High-Resolution (UHR) Cone-Beam Computed Tomography (CBCT)

Sunghoon Choi [1], Chang-Woo Seo [2] and Bo Kyung Cha [3,*]

[1] Safety Measurement Institute, Korea Research Institute of Standards and Science (KRISS), Daejeon 34113, Korea; reschoi127@gmail.com

[2] Department of Radiation Convergence Engineering, Yonsei University, Wonju 26493, Korea; cwseo@yonsei.ac.kr

[3] Electro-Medical Device Research Center, Korea Electrotechnology Research Institute, Ansan 15588, Korea

* Correspondence: bkcha@keri.re.kr

Received: 22 September 2020; Accepted: 6 November 2020; Published: 10 November 2020

Abstract: In this study, the effect of filter schemes on several low-contrast materials was compared using standard and ultra-high-resolution (UHR) cone-beam computed tomography (CBCT) imaging. The performance of the UHR-CBCT was quantified by measuring the modulation transfer function (MTF) and the noise power spectrum (NPS). The MTF was measured at the radial location around the cylindrical phantom, whereas the NPS was measured in the eight different homogeneous regions of interest. Six different filter schemes were designed and implemented in the CT sinogram from each imaging configuration. The experimental results indicated that the filter with smaller smoothing window preserved the MTF up to the highest spatial frequency, but larger NPS. In addition, the UHR imaging protocol provided 1.77 times better spatial resolution than the standard acquisition by comparing the specific spatial frequency (f_{50}) under the same conditions. The f_{50}s with the flat-top window in UHR mode was 1.86, 0.94, 2.52, 2.05, and 1.86 lp/mm for Polyethylene (Material 1, M1), Polystyrene (M2), Nylon (M3), Acrylic (M4), and Polycarbonate (M5), respectively. The smoothing window in the UHR protocol showed a clearer performance in the MTF according to the low-contrast objects, showing agreement with the relative contrast of materials in order of M3, M4, M1, M5, and M2. In conclusion, although the UHR-CBCT showed the disadvantages of acquisition time and radiation dose, it could provide greater spatial resolution with smaller noise property compared to standard imaging; moreover, the optimal window function should be considered in advance for the best UHR performance.

Keywords: ultra-high resolution; cone-beam computed tomography; low-contrast object; optimal filter; modulation transfer function; noise power spectrum

1. Introduction

Ultra-high-resolution (UHR) computed tomography (CT) has been used in commercial applications since 2017 due to it features of higher image spatial resolution [1]. Kakinuma et al. used a prototype UHR-CT that was operated with 0.25 mm detector pixel size and 0.1 mm reconstructed image pixel interval at 0.25 mm slice thickness [2]. Another experiment with a clinical UHR-CT scanner (Aquilion Precision, Canon Medical Systems) reported that the system was operated in the same condition of that in the previous paper [3]. The clinical UHR-CT has three scan modes: normal, high-resolution (HR), and super-high-resolution (SHR) modes, which support 512 × 512, 1024 × 1024, and 2048 × 2048 image matrixes, respectively, in a given reconstruction field of view

(FOV) [4]. Judging from the previously published papers, UHR-CT generally can be distinguished from conventional high-resolution CT (CHR-CT) because it makes use of pixel sizes below 0.25 mm at image matrixes above 512. It has been reported that CHR-CT uses pixel sizes ranging from 0.23 mm [5] to 0.35 mm [6]. The clinical aspects of UHR-CT include reduction of vascular continuity of the coronary arteries, visualization of fine structures of lungs, such as peripheral pulmonary vessels less than 1 mm in size, and artifact reduction such as blooming [7,8].

Recent X-ray detector technology in both multi-row and the flat-panel detectors (FPDs) enables high-resolution acquisition at a small pixel size of less than 0.25 mm [9]. Following these efforts, dedicated UHR cone-beam CT (CBCT), e.g., the OnSight 3D system (Carestream Healthcare, Rochester, NY, USA), has been introduced for extremity scans at lower cost and radiation doses compared to multi-detector CT (MDCT) systems [10]. The OnSight 3D system are mounted with a CsI:Tl scintillator-based complementary metal-oxide semiconductor (CMOS) FPD with a pixel size of 139 μm. UHR-CBCT can ultimately improve the visualization of bone morphometry and contribute to the diagnosis of osteoporosis and osteoarthritis, and detection of fine fractures, which typically require measurements in the range of 0.05–0.2 mm [11]. In general, the FPD could be operated in detector pixel binning mode, which is the process of combining the adjacent electric charges into one pixel [12]. This can reduce both the electronic and quantum noise, and decrease the image readout time at a higher frame rate. The user selects the FPD operation in either full or binning mode, which can optimally satisfy the need of correlation between the resolution and frame rate.

However, the image at higher resolution is not always good, especially for low-contrast detection tasks due to the enhanced noise level during the process of filtered back-projection (FBP) image reconstruction [13]. The "low contrast" of the image can be described as low discrimination between the target and background. The spatial resolution measurement in high-density materials, such as bar pattern and tungsten wire, is an easy task for both standard and UHR CT imaging. However, medical image quality of low-contrast objects is defined in terms of how well the tradeoff relationship between the resolution and noise is obtained from the image [14]. The amount of noise suppression at high frequencies is adjustable by setting either different cutoff frequency levels or different smoothing functions implemented on the CT sinogram. The higher the cutoff frequency level, the sharper but noisier the reconstructed image [15]. This, in turn, results in reconstructed image quality, thereby greatly influencing the detectability of objects by human observers [16]. Unfortunately, choosing an optimal filter scheme relies on experience, because there is no global function that can accept all principal signals underlying the entire frequency range. Therefore, the effect of the reconstruction filters on different materials in UHR-CBCT should be studied to provide useful information when observing a tiny amount of information during UHR acquisition.

In this study, we measured the spatial resolution of five different cylindrical objects according to four different UHR acquisition modes using six different filter schemes. The self-developed UHR-CBCT system, which is installed at the authors' institution (Korea Electrotechnology Research Institute, Ansan, Korea) was used for acquiring the CBCT images in both standard and high-resolution modes. This study aimed to evaluate the effect of filter schemes on the spatial resolution that underlies each imaging object and to suggest the optimal filter scheme in UHR-CBCT depending on the different object materials.

2. Materials and Methods

2.1. Ultra-High-Resolution Cone-Beam Computed Tomography System and Imaging Configurations

A photograph and a specification of a prototype CBCT system are provided in Figure 1 and Table 1. Our system was mounted with an amorphous silicon (aSi)-based thin-film transistor (TFT) array FPD (PaxScan 4030CB, Varian Imaging Products, Palo Alto, CA) and was operated in full and binning acquisition modes. As shown in Table 2, the imaging configuration was categorized into four subsections according to the two acquisition resolution setups and two reconstructed image resolutions. Each configuration was named depending on the row and column number of the matrix.

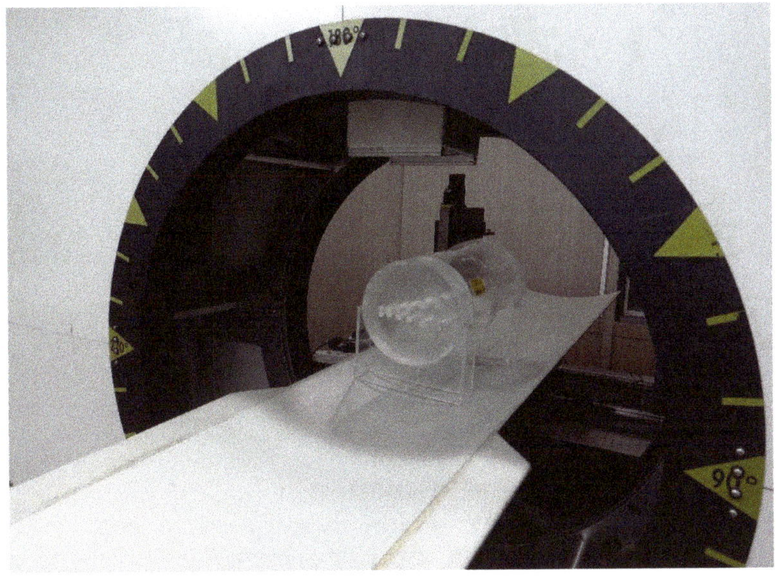

Figure 1. Photograph of the prototype cone-beam computed tomography (CBCT) system capable of both standard and ultra-high resolution (UHR) acquisition.

Table 1. Specifications of the imaging conditions.

Gantry	Sweep angle	0° to 360° with 1° step	
	Source-to-detector distance	1330 mm	
	Isocenter-to-detector distance	660 mm	
X-ray tube	Tube voltage	40–120 kVp	
	Tube current	10–500 mA	
	Exposure duration	16 ms	
		Standard acquisition	UHR acquisition
FPD	Image matrix	1024 × 768	2048 × 1536
	Pixel interval	0.388 mm	0.194 mm
	Framerate	7.5 fps	30 fps
	Readout time per view	~55 ms	~220 ms
	Total acquisition time	24 s	48 s
	Total entrance surface dose (ESD)	2.82 mGy	11.3 mGy
		Standard reconstruction	UHR reconstruction
Reconstruction	Image matrix	512 × 512	1024 × 1024
	Pixel interval	0.3 mm	0.15 mm

The center of rotation of the system was registered using the calibration phantom while rotating a full 360° with a 1° angle step for projection view image acquisition of 361 images. The 0.25 and

0.5 mm slice thicknesses were chosen based on previous studies [2–4]. The readout time of FPD with a 2×2 binning mode acquisition was four times faster than that of a full mode acquisition; therefore, a lower total acquisition time and lower radiation exposure were achievable owing to the higher framerate in the binning mode. All FBP reconstruction algorithms were self-programmed and coded in C++ with the CUDA toolkit version 10.0 using a single GPU card (GTX Titan-Xp, NVIDIA Co., Ltd., Santa Clara, CA, USA).

Table 2. Each configuration protocol with different resolution settings.

	Standard Reconstruction	**UHR Reconstruction**
Standard acquisition	Configuration (1, 1)	Configuration (1, 2)
UHR acquisition	Configuration (2, 1)	Configuration (2, 2)

2.2. CT Performance Phantom

We used the CIRS Model 610 American Association of Physicists in Medicine (AAPM) CT performance phantom to measure spatial resolution and noise property. The CT number linearity insert (Part No. 610-02), which includes five cylinders with different densities, was a targeted imaging object for resolution measurement. The detailed specifications of the inserted cylinders are given in Table 3. Each cylinder has the same size and shape and has a low contrast against the background material, thus presenting a small absolute signal difference between the two materials. The larger the material index, the smaller the difference between the background and target material densities. Note that a small absolute difference between the densities of two materials does not always guarantee a small image contrast because the CT numbers are represented by the linear attenuation coefficients which are dependent on both X-ray energy and density.

Table 3. Material index and name of each cylinder embedded in the American Association of Physicists in Medicine (AAPM) phantom.

Material Index	**Material Name (Density (g/cc))**
M1	Polyethylene (0.95)
M2	Polystyrene (1.05)
M3	Nylon (1.10)
M4	Acrylic (1.19)
M5	Polycarbonate (1.20)
Background	PMMA * (1.18)

* Poly methyl methacrylate (PMMA).

The insert (Part No. 610-01-05) is comprised of a uniform material with an aluminum pin at the center, and is a good candidate for measuring noise power. We assumed that the noise behaviors were the same for all materials because quantum and electronic noise, which are both stochastic events, are dominant over the entire area.

2.3. Ramp Filter Design in Spatial Domain and Six Different Window Functions

Linear filtering can be categorized into two methods: applying the convolution kernel in the spatial domain and linear multiplication of a transfer function in the Fourier domain. A band-limited ramp filter constructed in the Fourier domain is defined as follows:

$$RAMP(\omega)_A = \begin{cases} |\omega|, & \text{if } |\omega| \leq 0.5 \text{ lp/mm} \\ 0, & \text{otherwise} \end{cases} \tag{1}$$

where ω is the discretized spatial frequency by considering the Nyquist frequency. However, the ramp filter in Equation (1) has a zero at $\omega = 0$ lp/mm such that the signals at the DC offset (zero frequency component) after linear multiplication go to zero. The Fourier transform of the ramp convolution kernel constructed in the spatial domain can be defined as follows [17]:

$$RAMP(\omega)_B = FT\{ramp(n)\} = \int_{-\infty}^{\infty} ramp(n)e^{-i2\pi\omega}d\omega \qquad (2)$$

$$ramp(n) = \begin{cases} 1/4, & \text{if } n = 0 \\ 0, & \text{if n is an even number} \\ -1/(n\pi)^2, & \text{if n is an odd number} \end{cases} \qquad (3)$$

The ramp filter in Equation (2) does not include a zero, as shown in the comparison of the two shapes in Figure 2. Filtering with non-zero conditions avoids the zero signals that might have occurred if the filters were used with zero conditions.

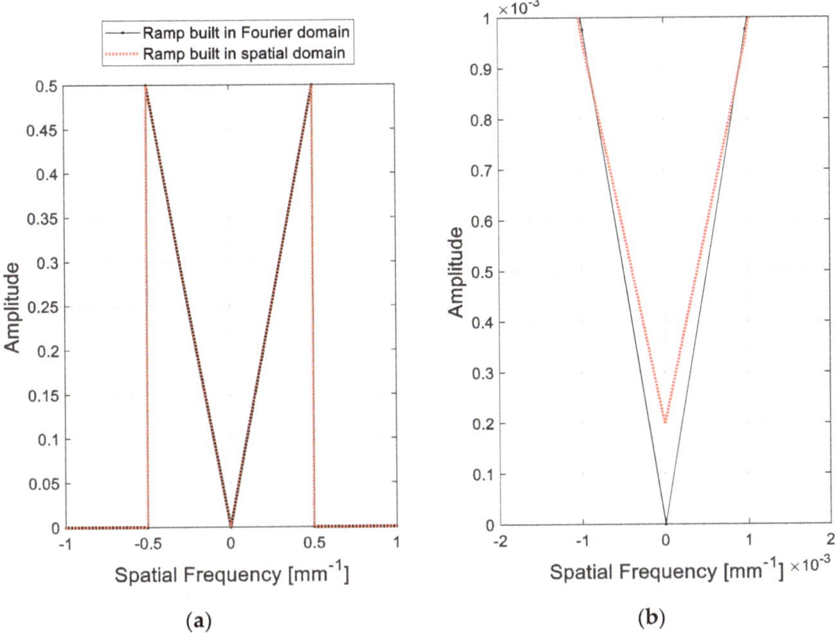

Figure 2. (**a**) Comparison of the ramp filters designed in different domains and (**b**) its magnified plot near the DC (zero frequency) component.

Many window functions have been introduced depending on the strength of noise suppression at different cutoff frequencies for each purpose [18]. However, the reduction of the critical signal is inevitable during noise suppression; therefore, the optimal window function is often heuristically chosen after multiple reconstruction trials. Six different smoothing windows were implemented herein in the ramp filter. Each window function was followed by the equation summarized in Table 4, where a term L in (b), (c), (d), and (f) indicates the length of the window.

2.4. Modulation Transfer Function (MTF)

Spatial resolution for each imaging configuration and each filter scheme was evaluated by the MTF measurement of the cylindrical materials as conducted by Richard et al. [19]. After subtracting the two-dimensional planar fit from the original region of interest (ROI) of each targeted cylinder, the radial pixel values around the edge of the circular shape were rearranged to yield a one-dimensional edge spread function (ESF). When converting the image grid from a Cartesian to polar map, the center of each disk was measured on a binary image through a gray-level threshold. The ESF, which is equivalent to the radial profile of the circle, was resampled with one-tenth of the reconstructed pixel size to reduce the non-uniformly distributed pixel noise [20]. The final ESF was derived by averaging the ESFs measured from consecutive axial slices. The MTF was the Fourier amplitude of the derivative of the ensemble-averaged ESF. In addition, the high-frequency noise of the ESF derivative was relieved through a Hanning window having the same length as the ESF size. The overall process of radial MTF measurement is depicted in Figure 3.

Table 4. Description of each window that was implemented with the ramp filter.

Window Title	Equation
(a) Butterworth A [1]	$1/\sqrt{1+\left(\frac{\omega}{f_c}\right)^{2p}}$ if
(b) Hanning	$0.5\left(1+\cos\left(\frac{2\pi\omega}{L}\right)\right)$ if
(c) Hamming	$0.54-0.46\cos\left(\frac{2\pi\omega}{L}\right)$ if
(d) Parzen	$\begin{cases} 1-6\left(\frac{\lvert\omega\rvert}{L/2}\right)^2+6\left(\frac{\lvert\omega\rvert}{L/2}\right)^3 & \text{if } 0 \le \lvert\omega\rvert \le (L-1)/4 \\ 2\left(1-\frac{\lvert\omega\rvert}{L/2}\right)^3 & \text{if } (L-1)/4 \le \lvert\omega\rvert \le (L-1)/2 \end{cases}$
(e) Butterworth B [2]	$1/\sqrt{1+\left(\frac{\omega}{f_c}\right)^{2p}}$
(f) Flat Top	$0.21-0.41\cos\left(\frac{2\pi\omega}{L-1}\right)+0.27\cos\left(\frac{4\pi\omega}{L-1}\right)-0.08\cos\left(\frac{6\pi\omega}{L-1}\right)+0.006\cos\left(\frac{8\pi\omega}{L-1}\right)$

[1] $p = 6$, fc $= 0.4$, [2] $p = 2$, fc $= 0.15$.

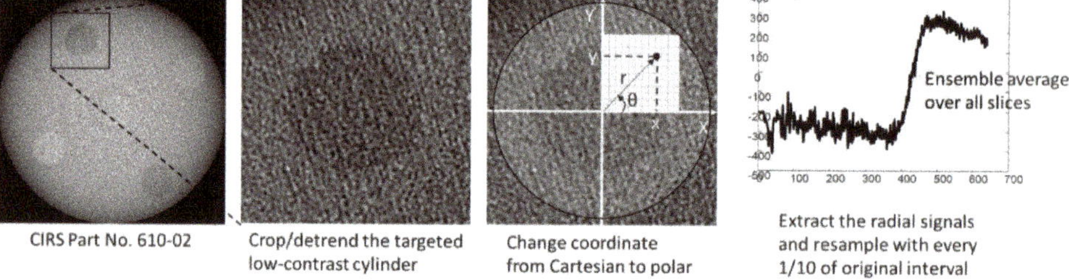

Figure 3. A depicted workflow for 1D edge spread function (ESF) measurement of the targeted low-contrast material.

2.5. Normalized Noise Power Spectrum

The normalized noise power spectrum (NNPS) was measured to quantify the noise level in the homogeneous volume of interest (VOI) of the poly methyl methacrylate (PMMA) background. The three-dimensional (3D) NPS was measured as described in Figure 4. The eight different VOIs without interference of any structure with the size of $150 \times 150 \times 45$ ($300 \times 300 \times 90$ for high-resolution reconstruction) were selected for measuring the 3D NPS. Each sub-volume overlapped with others to evaluate the radially and symmetrically distributed noise property (location independent noise pattern) [21].

Each mean subtracted sub-volume patch was Fourier transformed, absolute squared, and ensemble averaged to yield the power spectrum as follows [22]:

$$NPS\left(f_x,\, f_y,\, f_z\right) = \frac{1}{2}\frac{d_x d_y d_z}{N_x N_y N_z}\left\langle \left| \mathcal{F}\left[S(i,j,k)-\overline{S}\right] \right|^2 \right\rangle, \tag{4}$$

where f_x, f_y, and f_z are spatial frequencies (mm^{-1}), d_x, d_y, and d_z are pixel sizes (mm), N_x, N_y, and N_z are the numbers of voxels in the sub-volume patch, $\mathcal{F}[\cdot]$ is the fast Fourier transform operator, and $S(i,j,k)$

and $\bar{S}$ indicate each voxel value and the mean intensity of the sub-volume patch, respectively. The 1D NNPS can be derived by radially averaging the 3D NPS [23].

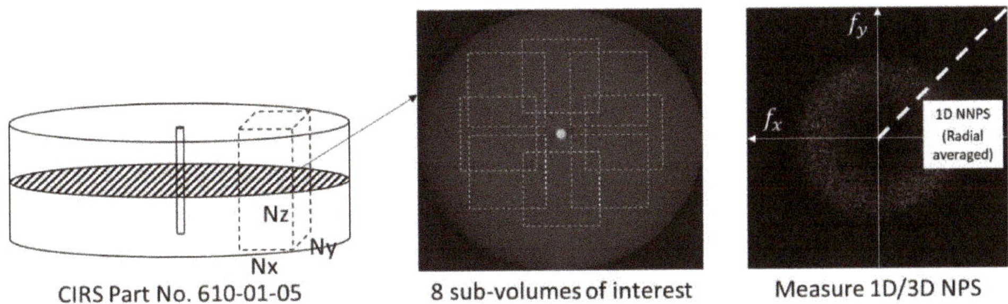

Figure 4. A schematic illustration for deriving 3D NPS.

3. Results

3.1. Filter Shape

Six different Fourier transformed and band-limited filters designed in the spatial domain with regard to the frequency response are depicted in Figure 5. Because the Fourier transformed sinograms were forced to be band limited with a band width of 0.5, the signals outside of the band frequency range went to zero, as shown in Figure 5. Similarly, each window function was also band limited and multiplied by the band-limited ramp filter. The magnitude of the filter at high frequencies was rejected when going from scheme (a) to (f) in Table 4, which is generally interpreted as noise suppression. Unlike other filters, some of the value of the flat-top window are negative.

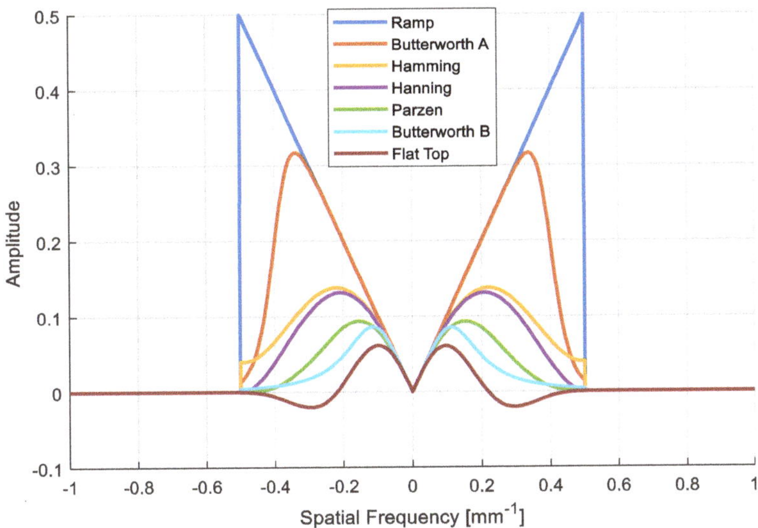

Figure 5. Different band-limited filter shapes as a function of frequency response.

3.2. Reconstructed Images with Different Filters and Configurations

Figure 6 shows the reconstructed images with configuration (1, 1) using the Hanning window and its cropped ROI images around the centers of five different materials. The relative contrast between each material and background with standard deviation error are plotted in Figure 6f. All five materials showed a low contrast, showing a small relative contrast below 0.15 (maximum contrast is 1).

As mentioned above, the higher density material does not always represent the higher CT number when we measure the contrast between each material and the background (PMMA). M2 and M5 showed the lowest contrast among the five materials.

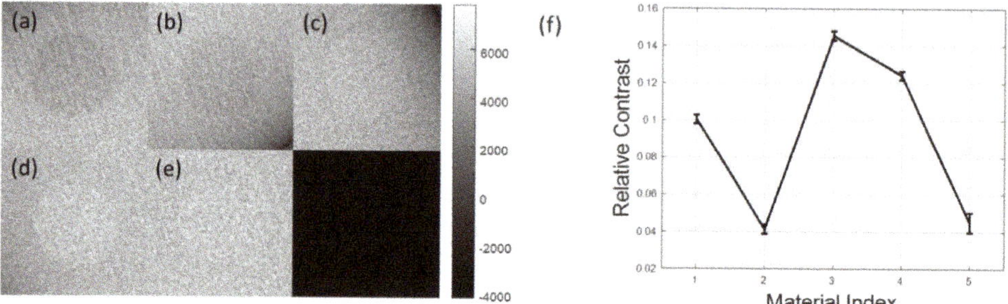

Figure 6. Reconstructed image with configuration (1, 1) using the Hanning window. (**a–e**) The cropped region of interest (ROI) around the center of each material, and (**f**) relative contrast between the target and background with standard deviation error bar as a function of each material.

The reconstructed cropped images of M1 with different configurations using the Hanning window are shown in Figure 7a. Figure 7b shows the radial profiles of each image grid in Figure 7a. The images reconstructed using standard detector resolution (configuration (1, 1) and (1, 2)) showed an unstable fluctuation in their radial profiles at the initial radial location. On the contrary, the images of configurations (2, 1) and (2, 2) showed relatively flat signals.

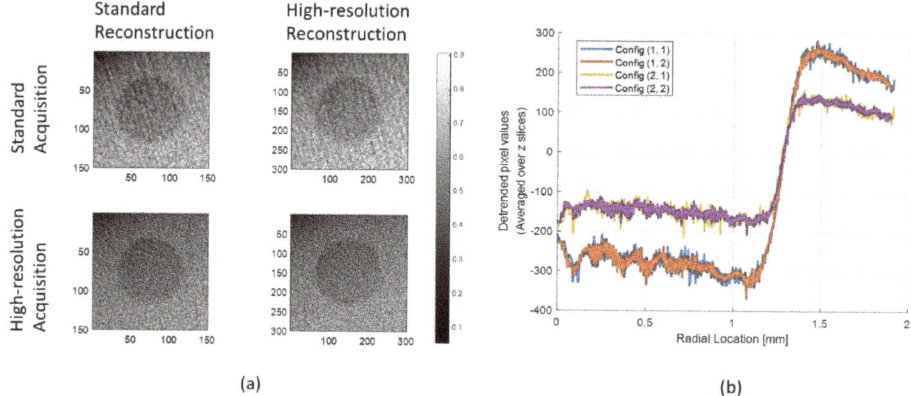

Figure 7. (**a**) Reconstructed image with different configurations using the Hanning window and (**b**) the radial profile of each configuration.

To understand the effect of filter schemes on image quality, the reconstructed images of M1 with different filters using configurations (1, 1) and (2, 2) are shown in Figure 8a. The radial profiles in Figure 8b correspond to the bottom row images in Figure 8a (configuration (2, 2)). The fluctuations of the radial profiles are gradually smoothed with an increase in the index number of filter schemes, demonstrating that the high-frequency noise was rejected by using the smoothing windows. The more oscillations in the signal, the coarser the MTF curve, as shown in Figure 9a.

3.3. Modulation Transfer Function

Six different MTFs for each filter scheme measured in the reconstructed images of M1 are shown in Figure 9. The higher the resolution of the reconstructed images, the better the MTF is preserved up to the high frequencies. In Figure 9f, f_{50}, which indicates the specific spatial frequency when the MTF is dropped to 0.5, was 1.39, 1.40, 2.52, and 2.57 lp/mm for the configurations (1, 1), (1, 2), (2, 1), and (2, 2), respectively. The effect of detector resolution on the reconstruction image resolution was minor when we compared the curves between configurations (1, 1) and (1, 2) (or (2, 1) and (2, 2)).

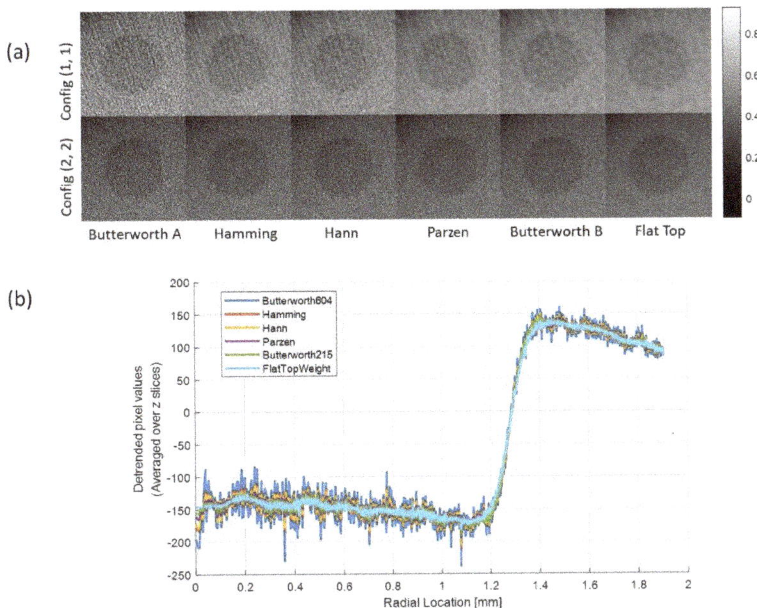

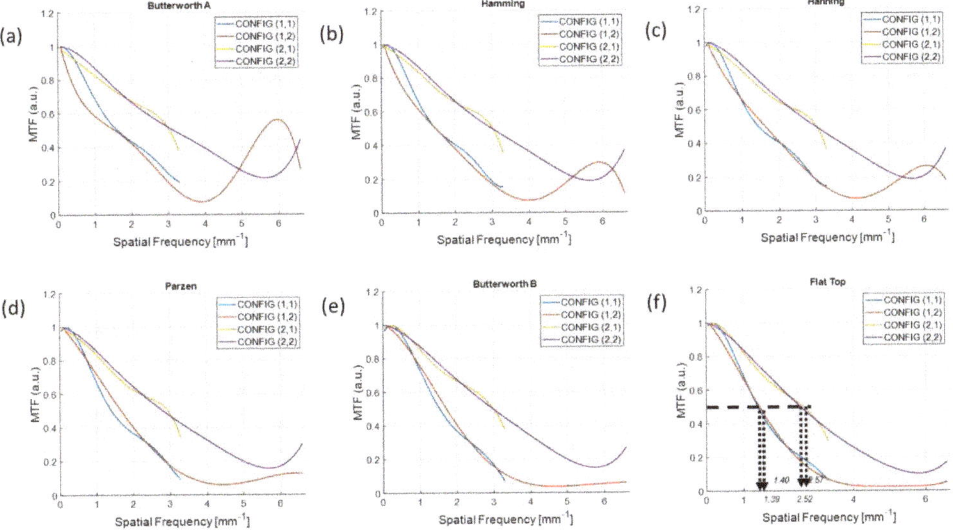

Figure 8. (**a**) Reconstructed image of M1 with different configurations according to the different windows and (**b**) the radial profile of configuration (2, 2).

Figure 9. Modulation transfer functions (MTFs) for different filter schemes from the (**a**) Butterworth A, (**b**) Hanning, (**c**) Hamming, (**d**), Parzen, (**e**) Butterworth B, and (**f**) Flat Top windows with different configurations. The f_{50}s measured in the images implemented with the flat top window were 1.39, 1.40, 2.52, and 2.57 lp/mm for the configuration (1, 1), (1, 2), (2, 1), and (2, 2), respectively.

The MTF curves measured in the reconstructed images of each material using configurations (1, 1) and (2, 2) are shown in Figures 10 and 11. As shown in Figure 11f, the f_{50}s were 0.94, 1.86, 2.05, and 2.52 and 1.86 lp/mm from M1 to M5, respectively, which demonstrates that MTFs were preserved up to high frequencies of the order of M3, M4, M1, M5, and M2; that is, in the order of the relative contrast in Figure 6f. In contrast, the imaging configuration (1, 1) not only did not follow the order of contrast, but also presented different orders of f_{50}s for the different filter schemes.

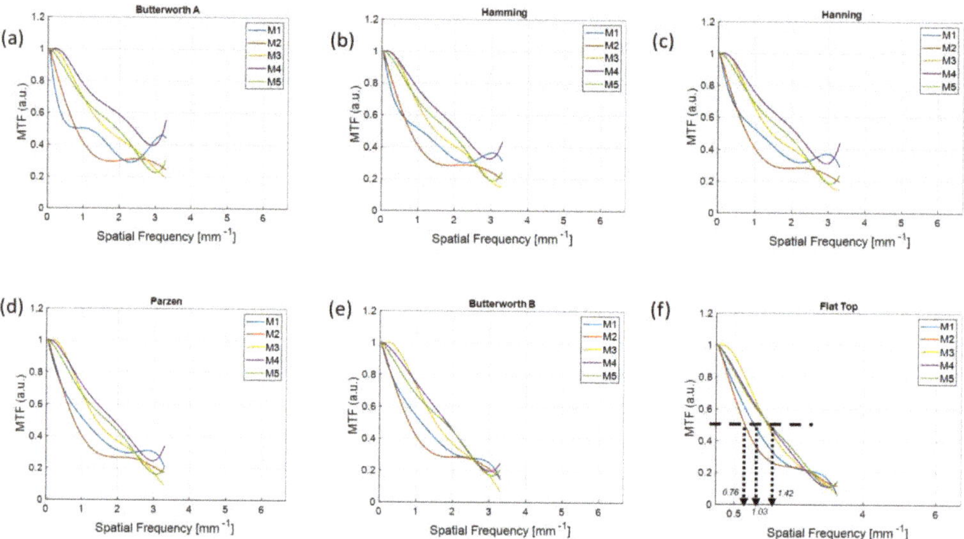

Figure 10. MTFs for different materials with the configurations (1, 1) using different filter schemes from the (**a**) Butterworth A, (**b**) Hanning, (**c**) Hamming, (**d**), Parzen, (**e**) Butterworth B, and (**f**) Flat Top windows. The orders of f_{50}s as a function of different materials were different for each filter scheme.

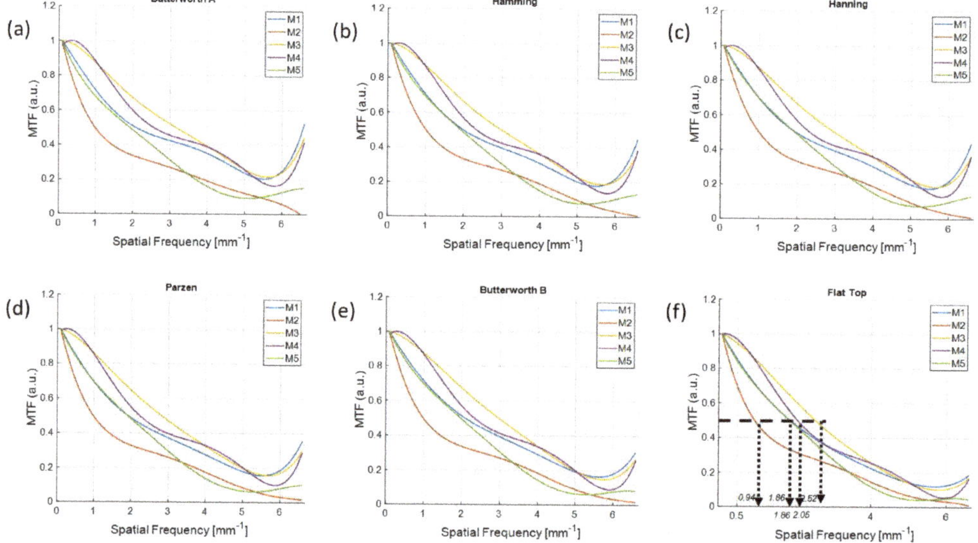

Figure 11. MTFs for different materials with the configurations (2, 2) using different filter schemes from the (**a**) Butterworth A, (**b**) Hanning, (**c**) Hamming, (**d**), Parzen, (**e**) Butterworth B, and (**f**) Flat Top windows. The order of f_{50}s as a function of different materials was M3, M4, M1, M5, and M2.

3.4. Normalized Noise Power Spectrum

Figure 12 shows the radially averaged 1D NPS for each configuration with different filter schemes. The standard reconstructed image resolution (configuration (1, 1) and (2, 1)) gave higher noise properties compared to the high-resolution images (configuration (1, 2) and (2, 2)). We also observed that the peak of the 1D NNPSs from the higher detector resolution was at larger spatial frequencies, which demonstrates that the noise was distributed up to a higher frequency when the smaller pixels were used in the detector. The NPSs decreased as the intensity of high-frequency smoothing increased.

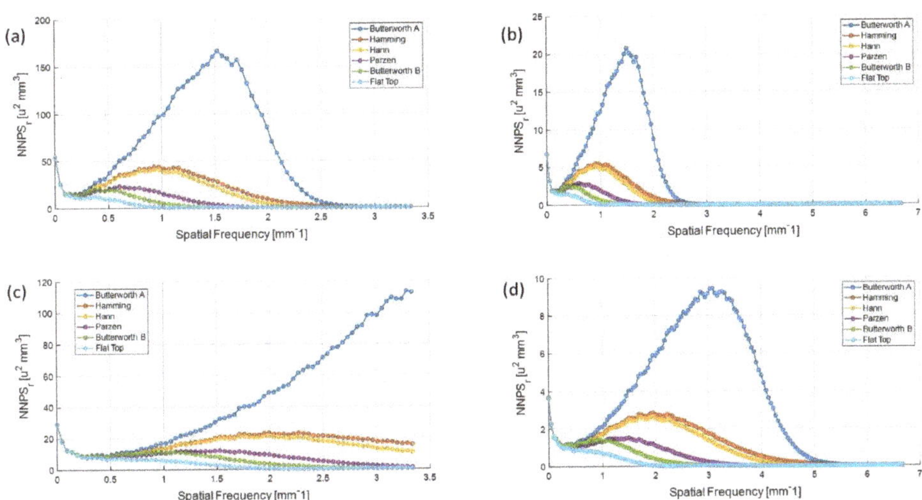

Figure 12. Radially averaged 1D NPS for each configuration from (**a**) (1, 1), (**b**) (1, 2), (**c**) (2, 1), and (**d**) (2, 2) with different filter schemes.

4. Discussion

We herein designed band-limited filters for all schemes. These can effectively retrieve the sampled projections because the projections are discretized into each detector pixel so that it is band limited in the Fourier domain [24]. As a result, band-limited filters lead to the removal of unnecessary noise signals at high frequencies.

There is no universal filter in CT imaging; therefore, the user should select an optimal smoothing window to observe the detailed internal structure with a purpose. Selecting an optimal window function is often based on experience rather than theory because we do not have a high level of knowledge about whether the imaging object is lying under a low-, mid-, or high-frequency range [25]. Thus, comparing the initial imaging performance of different filters and choosing the best solution for one's purpose is a good approach [25]. The most important factor when selecting the filter scheme is the manner in which the filter removes as many of the unnecessary components as possible in the frequency domain. In this experimental study, the signals near the edge of each material that we aimed to observe mostly lie in the low-frequency range, and show severe MTF distortion in the images applied with a high-pass filter, such as Butterworth A in Figure 9a. In contrast, the results in Figure 11f indicate that the flat-top window preserved the MTF up to a high frequency without an aliasing among the six filter schemes in our experiment. This is because the reconstructed images applied with the flat-top window not only resulted in uniform pixel values but also showed small oscillations (less noise) in both the target and background, as shown in the radial profiles in Figure 8b.

The flat-top window is used for cases in which a frequency component is required to be measured with great accuracy, e.g., a fixed-sine source [26]. Measuring the MTFs in the frequency domain could be interpreted as a discrimination of the signals spreading near the circular edge region. If a much larger signal difference exists between the target and the background, such as the tungsten edge, filter selection would not have been significant. However, we measured the MTFs for materials having no significant signal difference against the background material (low-contrast imaging); therefore, the amplitude accuracy was a key factor because the principal components in the Fourier domain were largely positioned in the low-frequency area [27].

The MTFs were preserved well at higher frequencies from the images reconstructed with a higher resolution. We observed that there was an MTF preservation loss up to 1.77 times by comparing the f_{50} between configurations (1, 1) and (2, 2) in Figures 10f and 11f when using the same target material and detector resolution. Therefore, using a UHR imaging protocol rather than a standard imaging configuration is recommended to understand the fine sharpness of low-contrast material if the detector is available to be operated at a higher resolution.

However, the high-level smoothing window is not recommended for standard resolution imaging configuration, as shown by the disagreement in the order of relative contrast in Figure 10. As shown in Figure 10, the flat-top window provided little difference in f_{50}s for different materials even though there was a clear discrimination in UHR imaging protocol. This was because the flat-top window overly smoothed the low-contrast object in the standard imaging, whereas the smoothing was still effective in UHR mode.

The trend of 1D NNPS in the configuration (2, 1) showed that the noise was distributed over all of the spatial frequencies. This demonstrates the back-projection from the high-resolution to small-image array would largely reduce the quantum noise and result in uniformly distributed noise.

The main drawback of this study is that all materials used to measure the MTFs had low contrast against the background PMMA intensity. This limits the study of higher-object-contrast materials such as bone and contrast-enhanced imaging. Our future study will be directed toward the effect of various filter setups on higher-object-contrast materials.

5. Conclusions

In summary, we observed the effect of filter schemes on several low-contrast materials using standard and UHR imaging protocols. Although UHR image acquisition requires a higher acquisition time and greater radiation exposure, we obtained spatial resolution up to 1.77 times higher than that of standard acquisition. In addition, the performance of UHR was affected by the FBP filter schemes, showing different f_{50} values and different noise patterns for different filters. Therefore, one should consider the optimal window function that can provide the best performance when observing the fine structure of the imaging object before UHR acquisition while comparing both the MTF and NPS.

Author Contributions: Conceptualization, S.C.; methodology, C.-W.S.; software, S.C. and C.-W.S.; validation, S.C.; formal analysis, S.C.; investigation, S.C.; resources, B.K.C.; data curation, S.C.; writing—original draft preparation, S.C.; writing—review and editing, S.C.; visualization, S.C.; supervision, B.K.C.; project administration, B.K.C.; funding acquisition, B.K.C. All authors have read and agreed to the published version of the manuscript.

Funding: This research received no external funding.

Acknowledgments: We would like to acknowledge the financial support from the R&D Convergence Program of NST (National Research Council of Science & Technology) of the Republic of Korea (CAP-15-04-KITECH).

Conflicts of Interest: The authors declare no conflict of interest.

References

1. Yoshioka, K.; Tanaka, R.; Takagi, H.; Ueyama, Y.; Kikuchi, K.; Chiba, T.; Arakita, K.; Schuijf, J.D.; Saito, Y. Ultra-high-resolution CT angiography of the artery of Adamkiewicz: A feasibility study. *Neuroradiology* **2018**, *60*, 109–115. [CrossRef] [PubMed]
2. Kakinuma, R.; Moriyama, N.; Muramatsu, Y.; Gomi, S.; Suzuki, M.; Nagasawa, H.; Kusumoto, M.; Aso, T.; Muramatsu, Y.; Tsuchida, T.; et al. Ultra-High-Resolution Computed Tomography of the Lung: Image Quality of a Prototype Scanner. *PLoS ONE* **2015**, *10*, e0137165. [CrossRef] [PubMed]

3.   Akagi, M.; Nakamura, Y.; Higaki, T.; Narita, K.; Honda, Y.; Zhou, J.; Yu, Z.; Akino, N.; Awai, K. Deep learning reconstruction improves image quality of abdominal ultra-high-resolution CT. *Eur. Radiol.* **2019**, *29*, 6163–6171. [CrossRef] [PubMed]

4.   Hata, A.; Yanagawa, M.; Honda, O.; Kikuchi, N.; Miyata, T.; Tsukagoshi, S.; Uranishi, A.; Tomiyama, N. Effect of matrix size on the image quality of ultra-high-resolution CT of the lung: Comparison of 512×512, 1024×1024, and 2048×2048. *Acad. Radiol.* **2018**, *25*, 869–876. [CrossRef] [PubMed]

5.   Yanagawa, M.; Tomiyama, N.; Honda, O.; Kikuyama, A.; Sumikawa, H.; Inoue, A.; Tobino, K.; Koyama, M.; Kudo, M. Multidetector CT of the lung: Image quality with garnet-based detectors. *Radiology* **2010**, *255*, 944–954. [CrossRef] [PubMed]

6.   Tsukagoshi, S.; Ota, T.; Fujii, M.; Kazama, M.; Okumura, M.; Johkoh, T. Improvement of spatial resolution in the longitudinal direction for isotropic imaging in helical CT. *Phys. Med. Biol.* **2007**, *52*, 791–801. [CrossRef]

7.   Heutink, F.; Koch, V.; Verbist, B.; van der Woude, W.J.; Mylanus, E.; Huinck, W.; Sechopoulos, I.; Caballo, M. Multi-Scale deep learning framework for cochlea localization, segmentation and analysis on clinical ultra-high-resolution CT images. *Comput. Meth. Prog. Biomed.* **2020**, *191*, 105387. [CrossRef]

8.   Oostveen, L.J.; Boedeker, K.L.; Brink, M.; Prokop, M.; De Lange, F.; Sechopoulos, I. Physical evaluation of an ultra-high-resolution CT scanner. *Eur. Radiol.* **2020**, *30*, 2552–2560. [CrossRef]

9.   Gupta, R.; Grasruck, M.; Suess, C.; Bartling, S.H.; Schmidt, B.; Stierstorfer, K.; Popescu, S.; Brady, T.; Flohr, T. Ultra-high resolution flat-panel volume CT: Fundamental principles, design architecture, and system characterization. *Eur. Radiol.* **2006**, *16*, 1191–1205. [CrossRef]

10.   Sisniega, A.; Thawait, G.K.; Shakoor, D.; Siewerdsen, J.H.; Demehri, S.; Zbijewski, W. Motion compensation in extremity cone-beam computed tomography. *Skelet. Radiol.* **2019**, *48*, 1999–2007. [CrossRef]

11.   Cao, Q.; Sisniega, A.; Brehler, M.; Stayman, J.W.; Yorkston, J.; Siewerdsen, J.H.; Zbijewski, W. Modeling and evaluation of a high-resolution CMOS detector for cone-beam CT of the extremities. *Med. Phys.* **2018**, *45*, 114–130. [CrossRef] [PubMed]

12.   Zhao, B.; Zhao, W. Imaging performance of an amorphous selenium digital mammography detector in a breast tomosynthesis system. *Med. Phys.* **2008**, *35*, 1978–1987. [CrossRef] [PubMed]

13.   Euler, A.; Solomon, J.B.; Marin, D.; Nelson, R.C.; Samei, E. A third-generation adaptive statistical iterative reconstruction technique: Phantom study of image noise, spatial resolution, lesion detectability, and dose reduction potential. *Am. J. Roentgenol.* **2018**, *210*, 1301–1308. [CrossRef] [PubMed]

14.   Gang, G.J.; Tward, D.J.; Lee, J.; Siewerdsen, J.H. Anatomical background and generalized detectability in tomosynthesis and cone-beam CT. *Med. Phys.* **2010**, *37*, 1948–1965. [CrossRef] [PubMed]

15.   Pauwels, R.; Araki, K.; Siewerdsen, J.H.; Thongvigitmanee, S.S. Technical aspects of dental CBCT: State of the art. *Dentomaxillofac. Rad.* **2015**, *44*, 20140224. [CrossRef] [PubMed]

16.   Richard, S.; Siewerdsen, J.H. Comparison of model and human observer performance for detection and discrimination tasks using dual-energy X-ray images. *Med. Phys.* **2008**, *35*, 5043–5053. [CrossRef]

17.   Zeng, G.L. Revisit of the ramp filter. In Proceedings of the 2014 IEEE Nuclear Science Symposium and Medical Imaging Conference (NSS/MIC), Seattle, WA, USA, 8–15 November 2014.

18.   Srinivasan, K.; Mohammadi, M.; Shepherd, J. Investigation of effect of reconstruction filters on cone-beam computed tomography image quality. *Australas. Phys. Eng. Sci. Med.* **2014**, *37*, 607–614. [CrossRef]

19.   Richard, S.; Husarik, D.B.; Yadava, G.; Murphy, S.N.; Samei, E. Towards task-based assessment of CT performance: System and object MTF across different reconstruction algorithms. *Med. Phys.* **2012**, *39*, 4115–4122. [CrossRef]

20.   Choi, S.; Lee, H.; Lee, D.; Choi, S.; Lee, C.-L.; Kwon, W.; Shin, J.; Seo, C.-W.; Kim, H.-J. Development of a chest digital tomosynthesis R/F system and implementation of low-dose GPU-accelerated compressed sensing (CS) image reconstruction. *Med. Phys.* **2018**, *45*, 1871–1888. [CrossRef]

21.   Gang, G.J.; Lee, J.; Stayman, J.W.; Tward, D.J.; Zbijewski, W.; Prince, J.L.; Siewerdsen, J.H. Analysis of Fourier-domain task-based detectability index in tomosynthesis and cone-beam CT in relation to human observer performance. *Med. Phys.* **2011**, *38*, 1754–1768. [CrossRef]

22.   Gang, G.J.; Zbijewski, W.; Stayman, J.W.; Siewerdsen, J.H. Cascaded systems analysis of noise and detectability in dual-energy cone-beam CT. *Med. Phys.* **2012**, *39*, 5145–5156. [CrossRef] [PubMed]

23.   Prakash, P.; Zbijewski, W.; Gang, G.J.; Ding, Y.; Stayman, J.W.; Yorkston, J.; Carrino, J.A.; Siewerdsen, J.H. Task-based modeling and optimization of a cone-beam CT scanner for musculoskeletal imaging. *Med. Phys.* **2011**, *38*, 5612–5629. [CrossRef]

24. Leong, L.K.; Kruger, R.L.; O'Connor, M.K. A comparison of the uniformity requirements for SPECT image reconstruction using FBP and OSEM techniques. *J. Nucl. Med. Technol.* **2001**, *29*, 79–83. [PubMed]
25. Reljin, I.; Reljin, B.; Papic, V.D. Extremely flat-top windows for harmonic analysis. *IEEE Trans. Instrum. Meas.* **2007**, *56*, 1025–1041. [CrossRef]
26. Gilland, D.R.; Tsui, B.M.; McCartney, W.H.; Perry, J.R.; Berg, J. Determination of the optimum filter function for SPECT imaging. *J. Nucl. Med.* **1988**, *29*, 643–650. [PubMed]
27. Gade, S.; Herlufsen, H. Use of Weighting Functions in DFT/FFT analysis. In *Windows to FFT Analysis*; Brüel & Kjær Technical Review No. 3; Brüel & Kjær: Nærum, Denmark, 1987.

Publisher's Note: MDPI stays neutral with regard to jurisdictional claims in published maps and institutional affiliations.

MDPI
St. Alban-Anlage 66
4052 Basel
Switzerland
Tel. +41 61 683 77 34
Fax +41 61 302 89 18
www.mdpi.com

Sensors Editorial Office
E-mail: sensors@mdpi.com
www.mdpi.com/journal/sensors